The Birdlife of Rottnest Island

by

D.A. Saunders and C.P. de Rebeira

Illustrated by C.P. de Rebeira

First edition 1985.
Published by DAS amd CPdeR.
24 Victoria Street, Guildford.
Western Australia.
 ISBN 0 9593247 2 0

Printed for DAS and CPdeR by Frank Daniels Pty. Ltd.

Cover photographs:
Sacred Kingfisher (above left); aerial view looking south from Armstrong Bay; and Osprey on nest at North Point.

To Eleanor who loved Rottnest Island
and "saw the wild ducks".

The Birdlife of Rottnest Island

CHAPTER 1: Introduction

"The Birdlife of Rottnest Island"; yet another book about birds. There are many available, ranging from Field Guides costing around $20 telling how to identify birds, to glossy coffee table publications full of lovely paintings costing several hundred dollars. Every book should have an audience and a specific purpose. The aim of "The Birdlife of Rottnest Island" is education.

Western Australians are fortunate in having Rottnest Island, one of a chain of three islands lying off the coast of metropolitan Perth. Garden Island is the closest to the mainland and is now connected to the mainland by a causeway via Cape Peron. It is 1100 ha in area and owned by the Commonwealth of Australia. The Royal Australian Navy base H.M.A.S. Stirling in located on Garden Island and for that reason access is strictly limited with the public only allowed to visit it by boat during daylight.

Carnac is the next island in the chain. It is the smallest of the three (16 ha) and is 3 km north of Garden Island. It is an A-Class Reserve with a gazetted purpose of Conservation of Flora and Fauna. Being a conservation reserve all animals are protected and it is also the home of a thriving colony of the Tiger Snake *Notechis scutatus,* a fact which is well advertised and keeps most visitors off the island.

Rottnest Island is the largest of the three islands (1900 ha) and the most isolated (18 km from Fremantle and 14 km from Carnac Island). Despite this distance it is the most accessible because it is an A-Class Reserve whose gazetted purpose is Public Recreation. It is visited by up to 250,000 people every year and at peak periods there may be over 9500 people staying on the island.

Islands have long held a fascination for people and Rottnest Island is no exception. To many people, and Western Australians in particular, it is a magic island to which they go "to get away from it all and relax". The fact that legs and bicycles are the principal forms of transport means that locomotion is slow, allowing time to view the surroundings. As a result, many people, who during other periods would not have time to notice the wildlife, see and perhaps wonder about the animals they observe. Many of these animals are used to humans walking or riding close to them and it is possible to get a good look at some of the birds on the island, without needing binoculars to study them. On the mainland, most people would not be able to travel from their house to the shops and see the range of different species which they are likely to encounter on the island.

Many visitors to the island are not bird watchers but the sighting of some bird may interest them. This book provides information about the birds that occur on the island and gives some idea of what is known about their lifestyle. The island's major habitats are described, each of the 49 species of bird most commonly recorded on the island are illustrated and an account of their basic biology given. Anyone not familiar with identification of birds should be able to pick out the common birds of the island and find out something about their habits.

Obviously a book like this depends on the knowledge of many people other than ourselves. Authors whose works have been consulted are named in the text and the references given in the bibliography.

Although there are only 49 species of bird which occur regularly on Rottnest Island, there are many more which turn up in ones and twos, then disappear. These vagrants are not described in the text because of the uncertainty of their occurrence but a full list of all species known to have been recorded on the island is given in Appendix 1, together with a statement of their status.

Acknowledgements

We would like to thank the following for assistance during our work on the island: Mary Bremner, Peter Curry, Richard Curtis, Mary Dyer, Mike Dyer, Peter Graham, John Ingram, John Izzard, Ian Parker, Alma de Rebeira, Ian Rowley, Romayne Saunders, V Saunders, Geoff Shannon, Graeme Smith, Glen Storr, Frank Thorson, Ian White and Ron Wooller for providing us with data or assisting during field work; the University of Western Australia for accommodation; the Department of Fisheries and Wildlife for transport; the staff of the Rottnest Island Board, particularly Charlie Hansen, Graham Lilleyman, Graeme Maider and Trevor Mills; the staff of Kingstown Barracks, S/Sgt. M. Bernasconi and Cpl. I. Humphrys for assistance while on the island; Dick Schodde of the Australian National Wildlife Collection and Glen Storr and Ron Johnstone of the Western Australian Museum for loaning specimens to us; Graeme Smith for commenting on the manuscript; Rob Fleming for advice on illustrations; and Samantha Saunders for typing the manuscript. To these people and many more we are extremely grateful.

CHAPTER 2: Bird Habitats

Rottnest Island (Fig. 1) may be divided into 6 major habitats: coast; salt-lakes; freshwater and brackish swamps; woodlands; heath; and settlements. Each of these habitats has its characteristic suite of birds but some, such as the Silver Gull, Welcome Swallow and Raven may be found in all habitats while others like the Golden Whistler are restricted in their choice of habitat.

The Coast

The coast is comprised of sandy beaches, rocky headlands and cliffs, exposed reef platforms, offshore islets and stacks (Plates 1 and 2).

The beaches vary in degree of isolation, those further from the settlements being less disturbed by people than those closer to the settlements. The beaches also vary in amount of seaweed deposition and in degree of protection from the prevailing winds. Each of these factors affect some species of bird more than others. The Ruddy Turnstone, for example, likes to forage among weed banks and will tolerate people on the beach at the same time. Other species, like the Whimbrel are not so bold and trusting.

The exposed reef platforms are very important feeding areas for many of the coastal species like the Whimbrel and Reef Heron and the small tidal range means that these platforms are accessible for much of the time (Plate 1b). The offshore stacks and islets (Plates 2a and 2b) are mostly well isolated and the birds which breed on them are protected from disturbance, particularly by humans. These sites are used as breeding platforms by the Osprey, Little Pied Cormorant, Pied Cormorant and Rock Parrot.

Although 13 species are commonly seen around the coast, some of them may be encountered in other habitats. The Rock Parrot breeds on the offshore islets and stacks but often feeds and drinks around the salt-lakes and the settlement at Thomson Bay. The Crested Tern breeds on islets in the salt-lakes but feeds and rests around the coast, while the Wedge-tailed Shearwater is only found around the coast when it is breeding.

The Salt-lakes

There are more than 200 islands larger than 10 ha off the coast of Western Australia. Thirty-four of these exceed 1000 ha and the largest is Dirk Hartog Island (about 60,000 ha) off the mid-west coast. Rottnest Island is quite small in comparison (1900 ha) but it is unique in that it is the only one which possesses a salt-lake complex. Ten per cent of the area of the island is made up of a series of supersaline lakes which dominate the eastern half of the island (Fig. 2).

The salt-lakes were formed when a cave system collapsed and filled with sea water. This system was subsequently cut off from the sea and now consists of three deep lakes (Government House, Herschell and Serpentine) and four shallow lakes (Baghdad, Pink, Garden and Timperley). In addition there are two other small lakes which are dry during the summer (Negri and Sirius).

The level of water in all of the lakes changes with the seasons, resulting in various widths of gently sloping beaches being exposed. In summer these may be extensive, creating in some places, expansive exposed sand bars or spits (Plate 3a), all of which are suitable as feeding areas for birds. In winter when water levels are highest some of the lakes merge (Baghdad and Vincent, Government House and Serpentine, and Serpentine and Timperley), the exposed areas are reduced and in some areas the water reaches the samphire vegetation at the margins of the lakes (Plate 3b).

There are freshwater seepages around most of the lakes and these areas are important for those birds which must drink fresh water. The Rock Parrot and the the White-fronted Chat depend on these sources of water, as does the Mountain Duck which defends them as part of its brood territory. These seeps are part of the discharge of underground water from aquifers, which are also tapped to provide water for human use. Particular care needs to be taken to ensure that they are not used to the point where they cease to discharge into the lakes as surface fresh water is very scarce on the island.

The salt-lakes have been extensively studied by staff and students of the Zoology Department of the University of Western Australia and Don Edward gives an account of their ecology, while Phil Playford of the Geological Survey of Western Australia provides an account of the formation of the lakes in "Research on Rottnest Island" a booklet published by The Royal Society of Western Australia.

The salt-lakes attract several species of transequatorial migrant which breed in the Arctic Circle and spend the northern winter feeding around the lakes. Consequently these lakes are important in the conservation of several species of bird which are part of the avifauna of other countries.

Fourteen species occur on or around the salt-lakes.

The Freshwater and Brackish Swamps

In addition to the salt-lakes there are eight swamps in interdune depressions which are variably saline depending on time of year: two species of bird occur on them. The swamps are: Bickley, Rifle Range, Parakeet, Bulldozer, Salmon, Lighthouse and Barkers. The eighth, Aerodrome, was formed when the runway was constructed and the filling cut off the southern edge of Government House Lake. In the early 1970s, during a phase of road construction, an act of ecological vandalism resulted in the degradation of some of these swamps. Marl was extracted from Bickley, Salmon, Parakeet, Bulldozer and Lighthouse and all of these were deepened to the point where they permanently contain water. Salinity levels have increased markedly. All but Lighthouse Swamp have bare areas around them and unsightly heaps of marl beside them. Rifle Range, Barkers and Aerodrome Swamps are all ephemeral, being dry in late summer. Only Barkers Swamp is undisturbed; Rifle Range Swamp is in the middle of the golf course with extensive areas of mowed vegetation around it; while Aerodrome Swamp has the mowed runway verge along its north side.

The Woodlands

The vegetation of the island has changed extensively since the first visit by Europeans when it was believed to have been extensively covered with trees, mainly *Melaleuca lanceolata* (Ti-tree) and *Callitris preissii* (Rottnest Island Pine). By 1919 about 66% of the island was believed to have been covered by these species or *Acacia* (Wattle) scrub. By 1941 this was reduced to 23%, by 1956 to 18% and woodland now comprises about 8% of the island (Fig. 3). An account of the vegetation changes since 1822 is given by Pen and Green in an article in "Research on Rottnest Island" published by The Royal Society of Western Australia.

At present most of the *Callitris* have disappeared and the few remaining trees are dying. The remaining *Melaleuca* woodland is scattered in isolated patches across the eastern half of the island but there is a fine stand between Bickley Swamp and the lookout at Henrietta Rocks and another at Watson's Glade. There are also patches of *Acacia* scrub which become dense and relatively tall (when compared with the surrounding heath).

Re-afforestation has been attempted in the past and is now a policy actively pursued by the Rottnest Island Board. Much of the woodland on the island is a result of these activities and the plantations are scattered over the island east of Narrowneck. Unfortunately two of the main species planted; Tuart *Eucalyptus gomphocephala* and Coastal Moort *E. platypus,* are fairly "sterile" in terms of birdlife and they do not support the same diversity as the original woodland in which twelve species occur.

The Heath

Heath is a habitat type which includes several vegetation associations and the term is used loosely to cover both heath and low scrub which clothes most of the island. Different species of plant are more important in some areas than in others and much of the central part of the island is dominated by a prickly heath plant *Acanthocarpus preissii* (Plate 5b) which belongs to the lily family. Four species of bird are commonly seen in or over the heath.

The Settlements and Disturbed Areas

This is another loose category of habitat which covers such areas as the settlements at Thomson Bay, Kingstown Barracks, Geordie and Longreach Bays as well as the golf course and runway (Plate 6b); four species depend on these areas.

CHAPTER 3: The Birds

In the following sections the 49 species of bird which occur regularly on Rottnest Island are described and this description should be read in conjunction with the plate which illustrates the bird. In the section on STATUS an account of the species is given together with a résumé of its biology. The length of these narratives varies greatly as very little is known about some species while other species have been extensively studied. Anyone wishing to know more about any particular species should look at the references given in the bibliography.

Red-capped Dotterel

SECTION I: Birds of the Coast

Pied Oystercatcher

WEDGE-TAILED SHEARWATER *Puffinus pacificus* (Plate 7).

Identification: Length 37-46 cm; weight 220-420 g. The sexes are similar with all of the feathered surfaces being brownish-black. The bill is dark grey, the eyes are brown, the feet flesh coloured and the tail wedge-shaped.

Status: The Wedge-tailed Shearwater is a common, regular breeding migrant which arrives in August. The birds breed in colonies which consist of burrows in soft sand or nests in crevices on the rocky, limestone islets. The two biggest colonies are on plateaux overlooking Cape Vlamingh and Radar Reef. In the summer of 1956/57 Glen Storr found 500 burrows at Cape Vlamingh; by September 1982 there were 940 burrows and 796 had fresh tracks around the entrance. These figures were obtained after the area had been soaked in a heavy downpour the day preceding the census. It would appear that this colony has nearly doubled since 1956/57. Storr also noted about 70 burrows overlooking Radar Reef. By September 1982 there were about 1630 burrows in this colony, a 23 fold increase during the intervening 26 years. Of this total, 1344 had fresh tracks around the entrance. In addition there was a small colony of about 60 burrows between these two large colonies. All three colonies consist of burrows in the sand among vegetation of the mixed succulent mat community. This community consists of *Rhagodia baccata, Threlkeldia diffusa, Atriplex cinerea, Frankenia pauciflora* and *Mesembryanthemum crystallinum.* There are several other colonies consisting of less than 20 nest sites around the main island. These are on Green, Dyer and Parakeet Islands, and possibly Cathedral Rocks.

On their arrival toward the end of August the birds excavate the burrow, which may have become clogged during the winter. When complete the burrow may be 1-2 m long, 15 cm wide and the terminal nesting chamber 30 cm in diameter. The excavation is done at night and nothing is known about the respective roles of male and female in this task. The single white egg is laid in November, hatching in January. The nestling leaves the burrow in April and the colony is empty by early May.

During the period between hatching and fledging the nestling is fed by its parents every few days. The parents start flying into the colony just after last light and the birds continue to arrive for several hours. These are the dark birds which the fishermen, at West End see at night, arriving and disappearing down burrows. As the birds enter the burrows they give a low moaning call, which is characteristic of this species. This is the common "mutton-bird" of the west coast of Australia. It is also found on the east coast and is distributed throughout the tropical and subtropical waters of the Indian and Pacific Oceans.

LITTLE PIED CORMORANT *Phalacrocorax melanoleucos* (Plate 7).

Identification: Length 50-65 cm; weight 560-970 g with the males being slightly larger than the females. This is the smallest of the Australian cormorants and is about half the weight of the Pied Cormorant, which also occurs on Rottnest Island. The Little Pied Cormorant is black on top, white underneath with white thighs. The bill is short in relation to the head (unlike the Pied Cormorant) and is predominantly yellow. The eyes are brown and legs and feet are black.

Status: In the late 1950s and early 1960s this species was an uncommon visitor and then only to Thomson Bay, despite the fact that it was not uncommon on the adjacent mainland. Now it appears to be a resident breeding in small numbers. It has been recorded from the coast all round the island usually in groups of less than four. The largest number known to have been present was 22 in June and July 1983. This species has been recorded on each of more than 30 visits we have made to the island since 1981.

Between July and October 1983 there was a small nesting colony of at least three breeding pairs on the south-west side of Duck Rock opposite Bathurst Point. The colony was observed through a telescope from Bathurst Point and each breeding adult was sitting on a platform of sticks. On 12 October a large fluffy chick was seen sitting on one of the nesting platforms.

This species usually lays 3-5 eggs, which are incubated by both parents. It will breed at any time of year if there is a suitable supply of food. In common with the other cormorants, it feeds on fish and crustaceans and is able to hunt for prey underwater. After swimming it may be seen standing with its wings spread out to dry.

PIED CORMORANT *Phalacrocorax varius* (Plate 7).

Identification: Length 75-89 cm; weight 1.4-2.5 kg with the males being larger than the females. This species is black on the upper surfaces, white on the undersides and has black thighs. The bill is long, grey or dark bone and there is a distinctive orange-yellow patch of skin between the eyes and the bill. The eyes are green and the legs and feet black.

Status: This is the most common cormorant found round the coast. It may be found singly or in small groups either swimming or perched on rock outcrops, beaches, boats, mooring buoys, etc. Large groups may be found on Phillip Rock, the beach at Phillip Point or on Dyer Island.

The only known nesting colony around Rottnest Island is on Dyer Island. On 16 May 1949 Don Reid found approximately 40 nests on the island but some were empty. Those which had nestlings were all well advanced. Glen

Storr visited the colony in September 1956, after breeding had finished, and found about 120 nest platforms in a thicket of *Nitraria schoberi* on the upper south-western slopes. We visited the colony in October 1982 and there were 121 nest platforms so the colony was the same size as it was 26 years earlier. There were the remains of dead nestlings and broken eggs littered round the colony and several King Skinks *Egernia kingii* were seen. These skinks are plentiful on some of the islands around Rottnest Island where they feed on seabird eggs and small nestlings during the breeding season.

On Dyer Island the eggs appear to be laid in March and April and this autumn breeding season is characteristic of populations in the south-west of Western Australia. The Pied Cormorant usually lays 2 or 3 eggs, although it may produce up to 5. Cormorant colonies are vulnerable to disturbance during certain times of the breeding cycle. When the nestlings are large enough to move they will jump from the nesting platform if frightened and they may not be able to get back. As the nesting platforms are close together many nestlings may be affected by a disturbance with disastrous consequences to the colony.

The Pied Cormorant feeds primarily on fish but it does eat crustaceans as well. Like the Little Pied Cormorant, the Pied Cormorant may be often seen standing on a perch with its wings spread out to dry.

REEF HERON *Egretta sacra* (Plate 7).

Identification: Length 49-65 cm; weight 300-470 g. The Reef Heron occurs in two colour phases; dark and white. With only two exceptions it is the dark phase which has been recorded on Rottnest Island. This is uniform dark grey with a small amount of white on the throat. The bill is grey and the legs greenish.

Status: The Reef Heron is a resident but there are probably only one or two breeding pairs. They have been seen, usually singly, hunting on reefs all round the island. The birds have been seen on Dyer Island, below Jeannies Lookout, Mabel Cove, Fays Bay, Cathedral Rocks, Wilson Bay, Radar Reef, Strickland Bay, North Point, Fairbridge Bluff, Wallace Island, Parker Point, Fish Hook Bay, Salmon Point, Cape Vlamingh, Porpoise Bay, Little Armstrong Bay and once we saw one inland standing beside Lighthouse Swamp.

Only two nests have been recorded: one in Nancy Cove and one on Dyer Island. The Reef Heron usually lays 2-5 eggs which are laid in September on a pile of sticks on a ledge or low shrubs.

The Reef Heron is a very stealthy hunter, crouching low as it seeks its prey, striking quickly with its long bill when food is spotted. Fish forms the bulk of its diet but crustaceans are eaten as well.

OSPREY *Pandion haliaetus* (Plate 8).

Identification: Length 47-59 cm; weight 0.9-1.8 kg, with the female being slightly larger than the male. A large eagle shaped bird with a white head and neck and a well defined dark brown stripe through the eye and down the neck. The underside and feathered legs are white while the back is brown as are both surfaces of the wings and tail. The feet are pale flesh coloured, the bill is black and the eyes are yellow.

Status: The number of pairs breeding around the island has varied between two and four. Between 1961 and 1963 Holsworth studied this species and in each year there were breeding pairs at Salmon Bay, Salmon Point and Crayfish Rock. Between 1980 and 1984 the numbers varied between two and four pairs: two in 1980, 1981 and 1984 (although there may have been more as all sites were not monitored in those years); four in 1982; and three in 1983. During these breeding seasons the nests at North Point and Salmon Point were used at least four times, Parker Point at least three times and Crayfish Rock at least twice. Although there are usually at least six birds present around the island, numbers do increase just before and into the spring breeding season when over 20 birds may be present.

There are at least 12 nest sites, the bulk of them being on limestone stacks off the coast of the main island. [North Point (offshore stack); Crayfish Rock (stack); Rocky Bay (stack); Eagle Bay (main island); Cathedral Rocks (stack); east side of Wilson Bay (stack — nest no longer in existence); west side of Strickland Bay (stack); west side of Mary Cove (main island); east side of Salmon Bay (stack); Parker Point (three nests scattered round — main island); and Wallace Island (stack)]. The Osprey has also nested on the gallery of the Bathurst Point Lighthouse and on Dyer Island (1889). Many of these sites have been used many times in the past as the nests are built on prominent positions from which the birds can keep an eye on their surroundings.

The nest is composed of a mass of interwoven sticks and debris. The blue plastic wrapping tape used on crayfish bait packets appears to be a favourite material as do pieces of nylon rope. One nest we examined was constructed of sticks, rope fragments, matted seagrass, bits of beach- washed planks and one old rusty beer can (which was pushed into the structure). There were food items, a cuttlefish skeleton and the rib cage of a dead bird. These nests are added to each year and when unused, fall into disrepair and may be blown apart by the wind and water. The nest on the stack in Rocky Bay was first recorded in 1928 or 1929. At present it has not been used as a nest for some years although the birds do use old nest sites as feeding platforms, probably because they provide such superb vantage points. As the nests may be added to each year they become very large and may be several metres across the base and over one metre from the bottom to the lip. The 2-3 eggs are laid in a cup which may be 18-20 cm deep and about 50 cm in diameter. This depression may be lined with matted seagrass or some other material. The eggs are laid between August and October and there may be young in the nest until after Christmas.

Holsworth banded ten young birds between 1961 and 1963. One of these was found dead a year later at Lancelin 112 km north. Another was shot at Wilson Inlet near Denmark (384 km south-east), 4½ months later and a third was found dead 80 km north-east of Albany 6-7 months after banding. It seems that the immature birds disperse quite widely after leaving their nest.

It is not known if the adults pair for the life of one of the partners and use the same breeding site each year, nor is it known how long the birds live.

The Osprey feeds mainly on fish and it is a spectacular sight to see these magnificent birds plunge from the sky with their talons extended, penetrate the water with these powerful feet and fly off carrying a struggling fish. They have been seen feeding on other prey including Quokka *Setonix brachyurus* and a King skink. We have found the remains of several birds on their nests including Antarctic Prion *Pachyptila desolata,* Kerguelen Petrel *Pterodroma brevirostris,* Ring-necked Pheasant *Phasianus colchicus,* as well as the remains of one Quokka and part of a Sea lion pup *Neophoca cinereus.*

On one occasion an Osprey was seen circling low over a colony of breeding Fairy Terns *Sterna nereis* located beside the runway on the side of Government House Lake. The terns were agitated and flying round calling loudly as the Osprey circled above them. At one stage the Osprey came low through the circling birds and turned, flying at right angles to the runway and directly into the path of a passenger aircraft which was landing. The Osprey veered off about half a metre in front of one of the propellers. Undoubtedly the Osprey and the pilots both got a fright!

The Osprey may be seen flying over all parts of the island or perched on rocky stacks, cliffs, navigation boards, dead trees, the central lighthouse and on beaches.

PIED OYSTERCATCHER *Haematopus ostralegus* (Plate 9).

Identification: Length 42-52 cm, weight about 550 g. The black, white and red colouring of this bird makes it very distinctive. The head, breast, shoulders, wing and tail are all black, in contrast to the abdomen, rump and a bar on each wing which are white. The long stout bill and the eyes are red while the legs are pink.

Status: The occurrence of the Pied Oystercatcher has changed over the last 60 years. In the 1920s it was regarded as a rare visitor, but by the early 1960s it was resident in small numbers favouring the beaches, being rarely seen round the salt-lakes. Now this species is found on all of the beaches, round most of the salt-lakes and is often seen foraging in groups on the golf course.

When foraging in this situation, away from the breeding territories, it may be seen in groups of up to 8 birds. Usually the Pied Oystercatcher occurs singly, in pairs or in family groups of 2 adults and up to 2 immatures.

The Pied Oystercatcher is territorial and there appears to be a resident pair on most of the beaches around the coast with between 18 and 25 breeding pairs scattered round the island. They nest above the high tide level, or on the cliffs overlooking their territorial beach. They have started to move off the beaches and now nest around the salt-lakes and in 1984 one pair nested on the exposed ground beside the sewerage tanks located on the road to the Basin.

The nest is a scrape in the ground in which 2 eggs are laid sometime between July and November. When the eggs hatch the chicks are led to the beach to forage with their parents.

When faced with any threat the adults get very agitated and the chicks scurry for cover and 'freeze', lying flat on the ground. The chicks rely on their cryptic colouration to render them invisible. The parents may give a 'broken wing' display to distract any intruder and lead it away from the chicks. Once this has been achieved the adults may proceed to harass the intruder by flying at it, calling loudly, until it leaves the area.

Our observations of a distinctly marked female (she had a malformed leg) indicate that the birds tend to maintain the same territories each year. She was observed over several seasons along the beach at Salmon Bay East. She was also seen foraging around the east end of Herschell Lake, indicating that birds breeding on the coast do forage in the interior of the island. It is likely that the pairs remain together throughout the year as their chicks remain with them until just before the next egg laying period. We banded a small 'runner' with its parents at Eagle Bay in early December when it was several weeks old. This immature bird was still foraging with the adults at Eagle Bay over six months later.

The Pied Oystercatcher uses its stout bill to feed on molluscs, marine worms and snails. They often feed on the beach in the tide wash area, running down the beach as the surf retreats, looking for holes in the beach and plunging their bill into the sand, sometimes as far as its base.

WHIMBREL *Numenius phaeopus* (Plate 9).

Identification: Length 43-49 cm; weight 300-400 g. The Whimbrel is predominantly mottled dark brown and buff, the crown is brown with a light grey stripe down the middle and there is a light grey stripe above the eyes. The rump is whitish with light brown bars and most of the under surfaces are heavily marked with brown stripes. The abdomen is white and the bill is thin, dark brown, down-curved and long, being ⅔ of the total length from

the back of the head to the tip of the bill. The eyes are dark brown and the legs long and olive-grey.
Status: The Whimbrel is a transequatorial migrant which breeds in north-eastern Siberia and spends the non-breeding season in the Southern Hemisphere (Fig. 4). It is uncommon in the south-west of Western Australia but it is a regular visitor in small numbers (less than 5) to Rottnest Island, where it may be seen throughout the year. It frequents the exposed reefs where it usually forages by itself. It seems to be a shy bird, easily startled by humans and nearly 90% of our sightings were made on the western half of the island. Whimbrels occasionally visit the salt-lakes as we have 10 records from Government House, Serpentine, Baghdad and Herschell Lakes. The only sighting we have of a group of birds was of 4 resting in dry vegetation at the top of Bickley Beach.

The diet is varied and includes molluscs, crustaceans, worms, beetles, flies, moths and spiders. In Siberia they feed almost exclusively on berries, something not available to them round the reefs of Rottnest Island.

BAR-TAILED GODWIT *Limosa lapponica* (Plate 9).

Identification: Length 36-43 cm; weight 194-520 g, with the female being larger than the male. It is usually seen in eclipse plumage (non-breeding) and in this phase the body is pale brown above, there is a white stripe above each eye, the tail has brown and black bars, the under surfaces are very pale grey, the bill is long, distinctly upturned, with a pink base and dark tip and the long legs are black.

This species may be seen occasionally in partial or full breeding plumage, in which case the head and body are brick-red with dark brown streaks.
Status: A transequatorial migrant which breeds in eastern Siberia and northern Alaska (Fig. 4), the Bar-tailed Godwit is a seasonal visitor. This species is present in small numbers (less than 20) from mid-September through to April, although there may be the occasional immature birds overwintering. Little is known of their migration route, but an adult banded in Australia was recovered in South Korea. Some birds start to gain their breeding plumage before heading north and birds in this plumage phase have been seen on the island in early April.

Occurring in groups of up to 8 birds, the Bar-tailed Godwit is mostly found around the coast, foraging along the beaches, using its long bill to probe the sand in search of food, or feeding on the exposed reefs. Half of our coastal sightings have been made on the beach at Salmon Bay, particularly the east end, and the reef at Mabel Cove; the other half being made at nine other locations (both beach and reef). In addition, the birds have been seen several times around Baghdad and Government House Lakes.

The Bar-tailed Godwit feeds on amphipods, insects, molluscs, crustaceans and annelid worms.

SANDERLING *Calidris alba* (Plate 10).

Identification: Length 18-21 cm; weight 36-76 g. The Sanderling is the whitest of the sandpipers visiting Rottnest Island and in profile appears to be quite dumpy. The upper surfaces are pale grey and the brown centres of these feathers gives the upper surfaces a mottled appearance, the forehead is white, the under surfaces and the sides of the neck are pure white, the bill is straight and black, the eyes brown and the legs black.
Status: The Sanderling is a transequatorial migrant which breeds in the Arctic and spends the northern winter over a large part of the Southern Hemisphere, including the coasts of South America, Africa, Australia and New Zealand (Fig. 4).

While not common, it is a regular non-breeding visitor which arrives in late August and early September and leaves in April and May. Like the other migrating waders, numbers on the island vary between years. For example, there were no birds present during the winter of 1982, but about half a dozen remained through the winter of 1983. The largest number recorded was 110 in February 1982 and the number present changes as birds move about locally, visiting other beaches around Cockburn Sound.

This species is most commonly observed in flocks on the beaches where its feeding habits are quite distinctive. The birds chase after the receding waves picking up food, following the waves until they ebb, then turning and running up the beach in front of the incoming water. They continue this process until they have finished feeding, roosting high on the beach in some sheltered spot. The diet consists of small crustaceans, molluscs, worms, insects, amphipods and polychaetes.

Sanderling have been observed on all of the salt-lakes and Bickley Swamp, but they are less common there than on the beaches. Of the salt-lakes, Government House and Serpentine Lakes appear to be favoured with 62% of the inland observations being made there. Of the beaches, those along Salmon Bay are by far the most commonly visited by this species with nearly 70% of the 962 birds seen on the coast being recorded from there. One of the best vantage points for observing Sanderling is the spot where the sealed coast road overlooks the east end of Salmon Bay. This spot is often frequented by the birds, which roost high on the beach if humans are not present.

SILVER GULL *Larus novaehollandiae* (Plate 11).

Identification: Length 38-43 cm; weight 300-400 g. The Silver Gull, also called the Seagull, is the only gull occurring on Rottnest Island. The bird is white with pale grey upper surfaces and black wing tips. The eyes are white while the eye rings, the bill and legs are red. Immature Silver Gulls are mottled with brown and the bill and legs are black or grey depending on the age of the bird.

Status: The Silver Gull has been a common resident for as long as records have been published. Up until the early 1960s the breeding sites were scattered on most of the islets and stacks around the coast but there had been only one record of a nest on the main island; at King Head. Since then there has been a tremendous increase in the breeding range of the birds on the main island and nests have been recorded on Herschell Islet, the islet between Government House and Serpentine Lakes, Baghdad Islet, Garden Lake Islet, around Bathurst Point, east end of the Basin, Point Clune, the coast between Georgie and Little Parakeet Bays, Armstrong Beach, several places around the margins of Baghdad Lake, Barker Swamp, beside the Causeway, and City of York Bay. These sites are in addition to the former nesting sites which are still being used. Of these colonies, those on the islets in Herschell, Baghdad and between Government House and Serpentine Lakes are the biggest on the island; the latter colony had 76 nests sites either ready for eggs or in use during late May 1984.

The Silver Gull has been studied in some detail in two studies: one by Lexie Nicholls, and the other by Ron Wooller and Nic Dunlop. Those investigations were conducted on the population breeding on Carnac Island but the results explain the breeding biology of the birds using Rottnest Island.

Once they reach sexual maturity the birds maintain permanent pair bonds and they also tend to return to the same nest site to breed each season. The nest is a depression on the ground and nests vary in the amount of vegetation in or around them. On Crayfish Rock the nests consist of a few bits of weed on thin layers of sand over rock; on the peninsula separating Vincent from West Baghdad Lakes the nests are scrapes in the shell grit, while on the islet between Government House and Serpentine Lakes they nest among the vegetation.

The clutch consists of 1-3 eggs and of 501 clutches examined on Rottnest Island, 31% were of 1 egg; 63% were of 2 eggs and only 6% of 3 eggs. The breeding season is extended with eggs being produced any time between April and November. There tend to be peaks of egg laying in the early part of the breeding season, in early winter and late August/early September. Breeding activitiy ceases during the summer.

The Silver Gull in the south-west of Western Australia is an interesting species because it is the only seabird known to be capable of rearing two broods of chicks each year. This potential for high reproductive success is offset by a high mortality of eggs and chicks. Lexie Nicholls found that during one breeding season 50 marked pairs of gulls produced 73 free flying young from 251 eggs: a success rate of 29%. Many of these free flying young would have died before reaching sexual maturity as the early stages before and just after reaching independence are particularly dangerous. In some species

80-90% of the young which leave the nest, die in the first six months of life. Using the figure of 80% mortality between leaving the nest and the age of six months: these 50 pairs would have produced 15 young.

This apparent low success rate for breeding is countered by the potential longevity in this species. Band recoveries have shown that the birds may live for a relatively long time. For example a bird banded on Carnac Island was found dead at Fremantle nearly 12 years after banding. Other band recoveries have indicated lives of over 18 years, enough time for the birds to replace themselves and provide the potential for expanding the population.

Some individuals are more experienced than others and are more successful in producing young. This is shown by some data collected by Nicholls on the 50 pairs she studied. Eighteen of these pairs were successful in double brooding in one season and they raised 41 free flying young from 83 eggs (49% success) compared with the remaining 32 pairs who only reared 32 young from 168 eggs (19%). Put another way, the 18 pairs comprised 36% of the sample of 50 pairs studied, yet they contributed 56% of the free flying young. Age and experience probably accounts for much of this difference.

This potential for expanding their population has been realized over the last 30-40 years and has resulted in a marked increase in the population of the Silver Gull around centres of human population. On Rottnest Island the birds are concentrated on the eastern third of the island, east of Pink Lake. They feed in large numbers on the rubbish tip west of Herschell Lake and also forage around the settlements where they quickly congregate around any person eating out of doors. Studies by John Iveson have shown that about 8% of Silver Gulls carry *Salmonella.* Salmonellosis is a common cause of gastro-enteritis in humans and occurs when contaminated material is eaten. The source of contamination is invariably animal droppings, particularly those of the Quokka, but any animal carrying *Salmonella* does pose some problems. With less than 1 in 10 gulls carrying infection there seems to be little cause for concern; unfortunately there are often 20-30 gulls hovering around a picnic table and the birds are not particular about where they leave their droppings. If people are not scrupulously clean in their approach to eating on the island the risk of infection is there. One of the delights of the island is the close relationship between humans and other animals, but the gull population is too high and bolstered by artificial food sources: handouts from visitors and the rubbish tip. These sources should be eliminated to bring the numbers of gulls down.

The Silver Gull is an opportunistic feeder and it eats a wide range of food. Among other items are eggs and young of several species of bird. We have seen them killing and eating Mountain Ducklings, Avocet and Fairy Tern chicks. The gulls often hang round the colonies of the Fairy Tern and when the nesting birds are disturbed and leave their nests, the gulls move in and steal eggs and chicks. As there is some concern about the status of the Fairy

Tern, the added pressure posed by the gulls should be reduced by reducing the numbers of gulls.

The gulls also eat the brine shrimp *Artemia sp.* which occurs in large numbers in the more saline of the salt-lakes. The birds feed on the surface of the lake in a similar manner to the Banded Stilt. They also feed on swarming ants. In early June 1982 the island was literally blanketed by swarms of ants flying off to establish new colonies. Gulls were spread the length and breadth of the island hawking for ants, behaving like swallows as they flew round snatching prey out of the air. The pilot of the commuter service to the island remarked that from above, the white gulls appeared to be distributed in a regular pattern all over the island.

BRIDLED TERN *Sterna anaethetus* (Plate 11).

Identification: Length 34-42 cm; weight 70-137 g. The back and wings are brown-black, there is a black cap which extends down the back of the neck, the under surfaces are dirty white and there is a narrow white forehead stripe which extends above and well behind each eye. The eyes are brown, the bill and legs are black and the tail is very deeply forked. This tern is sometimes known as the Dog Tern because its call is not unlike the barking of a puppy.
Status: The Bridled Tern is a common breeding visitor but little is known about the movements of the Bridled Tern once they leave the breeding areas. They do travel extensively as shown by one bird which was banded by Ron Johnstone of the W.A. Museum. He marked the bird on North Fisherman Island, north of Jurien Bay in early January 1973. It was recovered off the coast of north-east Borneo over 11 months later.

The breeding birds return to Rottnest Island in early October; in 1982 they had arrived by the 8th and in 1983 by the 10th. They lay one egg between late October and December, usually on a rock ledge near the water. Both sexes incubate and they are quite aggressive in defence of their nest, readily flying at intruders.

Their nesting distribution has changed over the last 25 years. Formerly they nested on the offshore islets and stacks including Dyer Island, Joan, Phillip, Duck, Monday and Clune Rocks. Now they nest around the coast of the main island as well as the islets and stacks. We have seen nests on Dyer and Parakeet Islands, Phillip, Clune and Monday Rocks and on rock ledges between the Basin and Pinky Beach as well as the headland between Geordie and Longreach Bays. We have also seen adults sitting or flying round other stacks and islets as well as rock ledges on the main island. This, together with the fact that they have been recorded as breeding on Cathedral Rocks, indicates that they now have a breeding distribution round much of the island, but not at present on Green Island.

The Bridled Tern feeds on fish and the birds remain until the young are free flying in late February or early March.

CRESTED TERN *Sterna bergii* (Plate 11).

Identification: Length 43-53 cm; weight 350-400 g. The Crested Tern, which is slightly larger than the Silver Gull, has a black crown and some of the feathers on the crown may be erected to form a crest. The upper surfaces are grey, the forehead, neck, chest and under surfaces are white. The eyes are brown, the long bill yellow and the legs black. In flight the wings appear long, narrow and raked back while the tail is deeply forked.

Status: The Crested Tern is the most common tern on the island and for many people, particularly those who stay at Geordie Bay, the calls of this bird may be one of the "sounds of Rottnest Island". The Crested Tern utters a harsh guttural cry as it flies and the nesting colonies on Baghdad and Herschell Lakes emit a constant racket at all hours of the day and night. The birds may be seen as they fly over heading to and from the coast. The breeding colonies produce so much noise that they may be heard outside the cottages at Georgie Bay (2km away) or the lighthouse on Wadgemup Hill (over 3km away).

The Crested Tern has altered the site of its breeding colonies over the last 30 years, but the population on the island is a subgroup of the much larger population which occurs on the islands of Cockburn Sound. There is an interchange between the various breeding colonies and a bird banded on Dyer Island was recovered at City Beach. During the 1950s this species nested on Dyer Island, Green Island and the stack in Eagle Bay but did not use each site every year. At this time the population was estimated at 300-400 breeding pairs and there were no records of any nesting on the main island. By the early 1980s the Crested Tern had moved its breeding colonies onto the salt-lake system. The two major breeding concentrations are now on the islet in West Baghdad Lake and the islet at the east end of Herschell Lake. The birds appear to have abandoned Dyer Island as a nest site but there was a small colony on Green Island early in 1982. It did not contain anywhere near the 300 nests recorded there in 1960.

The Crested Tern usually reaches sexual maturity when 2 years old but there are records of them having bred at 1 year. Once they attain maturity, they form lasting pair bonds but these occasionally break down and the birds remate. With the start of the breeding season the birds engage in courtship displays. During this period the members of the pair may be seen performing elaborate aerial displays as they dive, climb, circle and zig-zag in perfect close formation, calling loudly. On the ground during the prelude to copulation, the members of the pair engage in an elegant display as they circle each other with their wings held out from the body, drooped towards the ground and the crest raised.

The nest is a simple scrape in the ground and nests are packed tightly into the breeding colony; each one just beyond the reach of the neighbours' bill. The single egg (occasionally 2) is laid any time between April and November and incubated for about 4 weeks. During incubation and the nestling period the adults fly out to feed, bringing back fish to feed their partner or offspring. The fish are caught by plunge diving and seized just behind the head.

A common bird roosting on sand spits and shell beaches around the salt-lakes, the Crested Tern is also common on the beaches and rocky platforms around the coast. The sandy point between Stark and Rocky Bays is a favourite roosting point as is the east end of Salmon Bay. This is a particularly good vantage point to watch the birds as the road runs almost to the cliff and the birds may be observed from above without disturbance. At this spot the birds often roost in groups of several hundred; preening, sleeping, walking into the water and bathing, fishing just off the shore or involved in mating displays. The birds often roost on the rocky platforms at Cape Vlamingh and Cathedral Rocks where there are a series of rocky protrusions on which the birds perch, the area being well coated with white droppings.

Population counts of the Crested Tern on Rottnest Island carried out between March 1982 and October 1984 showed that there was a marked seasonal pattern in the number of birds present; being highest between April and June, depending on season (peak nearly 2200 in June 1983) and lowest in late winter and spring (about 620 in August and September 1983).

Banding studies have shown that the Crested Tern may live for more than 20 years.

ROCK PARROT *Neophema petrophila* (Plate 12).

Identification: Length 20-24 cm; weight 45-54 g. The upper surfaces are olive-brown, the under surfaces are yellow, there is a dark blue band across the forehead with pale blue surrounds, the flight feathers are dark blue, the eyes are brown, the beak is grey and the legs grey-brown.

Status: The Rock Parrot is the only resident species of parrot and it gave its name to Parakeet Bay, Little Parakeet Bay and Parakeet Island. It may be seen flying high and fast between the main island and the offshore islets, particularly Phillip and Clune Rocks. It was formerly much more common and a visitor in 1904 noted that it was fairly common but it appeared to favour the western side of the island. He also noted that it was breeding on several small islets and on the steep sides of some cliffs on the island itself. He remarked that he often disturbed pairs or small parties in the more open parts of the island. In the early 1930s another visitor recorded that the Rock Parrot was very common and found in flocks of many dozens. By the early 1960s the population had decreased to the point where Glen Storr noted that they were becoming scarce. The largest flock he saw was 21 in the samphire beside Lake Herschell in April 1955. Evidently there was considerable pressure on the population from aviculturists with young birds being taken from the nests and sent to the mainland for sale. Storr mentions that one island resident removed 30 young parrots from Parakeet Island (apparently in one breeding season). This had been going on for some years yet the resident could not understand why the birds stopped breeding there!

In the past the birds were known to have bred on Green Island as well as Parakeet Island and on Phillip and Duck Rocks. They certainly do not breed on Green or Parakeet now. The other two sites are relatively inaccessible and they may still breed there.

The birds select a crevice or hollow in the limestone as a nest site and lay 4-6 eggs sometime between August and October. The female incubates the eggs while the male forages, bringing food back to her several times a day and feeding her near the nest. Incubation lasts about 18 days and after hatching the nestlings are fed by both parents. The young birds leave the nest permanently about one month after hatching and join their parents in the foraging flock.

The population on Rottnest Island is isolated as this species does not occur on Garden Island or the adjacent mainland (except as a vagrant). Fortunately the illegal harvesting pressure seems to have been lifted and the population appears to be increasing gradually.

All of our 127 separate sightings have been made in the eastern half of the island and the birds are usually seen in ones or twos but we have seen several flocks of 12 and 13 and one flock of 15 feeding on low vegetation at the edge of the beach at Point Clune. The birds feed on samphire seed and the seed of Harefoot Grass *Lagurus ovatus* an introduced grass which is widespread in disturbed areas. It flowers in August and dies leaving a white ball of seed at the end of a short stalk.

The birds are not uncommon in the Thomson Bay settlement where they frequent the golf course and they have been seen at other more populous areas. For example, we saw one parrot feeding on Harefoot Grass beside the road, within 20 m of the Bakery at 10 o'clock in the morning, totally oblivious to people walking by. On another occasion we saw a bird feeding in an exotic pine tree between the Rottnest Island Board Office and the milk bar. The parrot was pulling seed from a mature pine cone which had opened but was still attached to the tree. Before the Army withdrew from Rottnest Island in December 1984, the staff at Kingstown Barracks told us that the birds often came to feed on the mowed areas around the barracks and readily came to sit under the sprinklers when the lawns were being watered. Because Rock Parrots are seed eaters they need to drink fresh water and they may be seen as they fly in to drink at some of the fresh water seepages around Government House and Baghdad Lakes, and Bickley Swamp.

SECTION II: Birds of the Salt-Lakes

Red-necked Avocet

WHITE-FACED HERON *Ardea novaehollandiae* (Plate 7).

Identification: Length 54-73 cm; weight 400-700 g. A blue-grey heron with the flight feathers being darker than those on the body. The area round the forehead, the eyes and throat is white, the legs greenish-yellow and the long bill is black.
Status: The White-faced Heron is an occasional non-breeding visitor and all of the reported sightings have been of lone birds with no apparent seasonality of occurrence. When present on the island the birds may be seen around the lakes, swamps or around the coast. They are known to feed on frogs, fishes, insects and crustaceans.

BLACK SWAN *Cygnus atratus* (Plate 8).

Identification: Length 90-140 cm; weight 4.1-6.4 kg. The body is black with the exception of the flight feathers which are white. The bill is red with a white bar near the tip. The legs and feet are black.
Status: Until the early 1960s the Black Swan had only been recorded once during the winter of 1961. Since then there have been varying numbers present; on occasions more than a dozen, but there are no published records of them having bred on the island. Between 1981 and 1984 there was usually only one bird but on several occasions a second was seen. Most of the sightings were around Garden Lake but a bird has been seen on the other lakes and swamps. The Black Swan eats plant material.

MOUNTAIN DUCK *Tadorna tadornoides* (Plate 8).

Identification: Length 62-70 cm; weight 1.0-2.1 kg. The Mountain Duck is the beautifully coloured duck which may be seen on or around all of the lakes and swamps and, at certain times of the year, around the beaches and cliffs of the island. The male is slightly larger than the female with a black head and neck. There is a white stripe around the base of the neck and a broad orange-brown band around the chest and shoulders. The remainder of the body is black except for a prominent white patch on the wings, a chestnut area behind the white and a vivid green diagonal bar below the white and chestnut patches. The feet and legs are grey-black and the bill is black. The female differs from the male in having a chestnut band around the chest and shoulders and a white ring around the eyes and the base of the bill. Immatures resemble the female but are duller and may have more white around the eyes.

Status: The Mountain Duck is the only resident species of duck. Between 1965 and 1974 it was the subject of a study by Tom Riggert of the Western Australian Department of Fisheries and Wildlife and it starred in the film 'Rottnest Winter', a superb wildlife film made by the Australian Broadcasting Corporation. This film detailed the private life of the Mountain Duck on the island and gave an insight into the problems of wildlife dealing with a world which had been modified by humans.

Although there is a resident population of ducks, numbers fluctuate with season as birds arrive from or depart for the mainland. The resident birds appear to be adults whch have formed pairs. Studies of individually marked animals have shown that this species will breed at the end of its second year and, once paired, form long-lived or semi-permanent bonds. These pairs usually remain together throughout the year although they do change mates occasionally.

The largest number of birds are usually present on the island in December and January, just after the breeding season and before the hot weather begins to cause the lakes to dry up and salinity levels to rise. Between 1965 and 1974 the largest number present on the island was 374 (January 1974) and there appears to have been no change in the period between, as 400 were recorded in December 1982.

As summer progresses, the lakes and swamps, which are swollen by winter rains, begin to evaporate. As they contract the ducks begin to congregate on several swamps or lakes and are very conspicuous as there are large groups of birds concentrated in small areas. The places which appear to be most favoured are Pink Lake, Garden Lake (and the lawns near the old boatshed on the east side of the lake), Bulldozer Swamp and the grassy edge of the west side of Lake Baghdad. At this time of year it is possible to see the entire population of the island in one group, as we saw in December 1981 when there were 391 birds on the lawns beside Garden Lake.

As summer progresses the birds begin to leave the congregations and go their various ways. The birds with strong pair bonds return to their brood territories around the lakes and swamps while the young birds and unpaired birds leave the island for the mainland. They may disperse widely over the south-west as birds from the island have been shot as far away as Esperance, over 600 km to the south-east. Some birds may remain in loose groups at this stage. These appear to be birds with weak pair bonds and those without mates. Some of these birds may try to break an established pair with the females trying to attract a mate. Unattached courting females are repulsed by the paired female whenever they approach the male.

During the summer the birds feed in the cool of the day in the salt-lakes and swamps, drinking at the fresh water seepages and resting during the heat of the day in groups or pairs. All through summer the birds continue to leave the island, then with the opening rains the behaviour of the birds changes. Those birds with established brood territories begin to defend them from other birds and a trip round the lakes and swamps will show pairs distributed around them with other birds trying to form pairs and seize a brood territory. Courtship behaviour is obvious, as is fighting between females over prospective mates. Pairs from the mainland begin to arrive to try to establish territories so the population increases slightly at this stage.

It is hard to overestimate the importance of a brood territory for the successful breeding of the Mountain Duck. This area must have a fresh water seep on it as it is this facility which allows the ducklings to feed in the super-saline lakes of the island. There are between 30 and 40 of these territories on the island and this places a limitation on the duck population.

As each pair establishes its brood territory (which is also a feeding area, including the adjacent lake) the birds defend it against other ducks. They must leave the territory to seek a suitable nest site and at this time of year they may be seen flying around, perched on headlands or cliffs, walking on beaches, or swimming in the ocean. The nest site is a crevice or cave in the limestone cliffs of the main island or offshore islets and stacks. In fact one stack off Bathurst Point is named Duck Rock because the Mountain Duck nests on it. The female does the prospecting for the nest site while the male stands guard, warning her of any threat. If disturbed, the female leaves the nest site and both birds fly off. Once a nest site has been chosen the female cleans it out and makes a depression in which to deposit the eggs. These nest sites may be used for many years and other species may also nest in them. One crevice in a cliff on the west end of Marjorie Bay has been used by the Nankeen Kestrel after the ducks vacated it.

Once the nest site is prepared the female commences egg laying. At this stage the pair leave the brood territory and fly to a point near the nest site. The female then flies to the entrance and goes in, while the male stands guard. After laying one egg the female leaves the nest and the pair fly back to the brood territory. This pattern continues (usually each day) until the clutch of 6-19 eggs is complete. Egg laying may take place between July and September, depending on seasonal conditions.

Incubation begins with the laying of the last egg in the clutch. The female lines the nest with down feathers plucked from her breast and covers the eggs with this insulating material when she flies to the brood territory to feed and drink (something she must do each day). The female does all the incubation which takes about 31 days. She is present at the brood territory for some time each day and when she returns to the nest she is escorted by the male. He does not land but circles as she enters the nest and he flies back to the brood territory.

As incubation commences with the laying of the last egg, the eggs hatch together and the ducklings leave the nest early on the second day after hatching. Both parents lead the brood by the most direct route to the brood territory. As the ducklings are not able to fly they must walk or swim all the way. As some birds nest at West End, this is a considerable journey for a 2 day old duckling to undertake (6 km to the nearest available territory). During the period from hatching to reaching the brood territory the ducklings live on their energy resources until they begin to feed themselves on the brood territory. The ABC film 'Rottnest Winter' showed a lovely sequence of a female leading her brood off a cliff, into the water and swimming out through the surf. During this travel to the territory the parents call continuously to help the ducklings along and the ducklings are vulnerable as Ravens, Silver Gulls and, according to fishermen, Pied Cormorants all prey on them. Feral Cats are also a problem as they are present on the island. In mid-July 1984 we watched a pair of Mountain Ducks leading a procession of 11 ducklings east down the road from West End. We saw the final stage of the journey as they walked by the edge of the main centre road past the Parker Point turn off, heading for Lighthouse Swamp. Both adult birds called as they walked and at one point the whole family scurried into the bushes because of some threat, the adults calling the ducklings out of the bushes when the danger had passed. They continued until they reached Lighthouse Swamp with the adults walking into the water and swimming into the centre. The ducklings followed with little hesitation. The ducklings learn to feed and drink by watching their parents and the family remain in the brood territory for one to two months. Both parents guard the ducklings and any threat will cause the adults to lead the family out into the comparative safety of the water. While the ducklings are small, Silver Gulls and Ravens pose the biggest threat to them. Gulls will swim between the adults and the ducklings to try and isolate one. As soon as they are able they will strike and carry it off. Ravens are a threat when the family is on land. We watched the family at Lighthouse Swamp deal with two Ravens which were threatening them after they arrived at the swamp. The male took off from the edge of the lake and flew at the Ravens calling loudly as the female led the ducklings for cover under some bushes. The male chased the Ravens several hundred metres away from the swamp, then circled and flew back to land near the point from which he had flown. The female brought the ducklings out of the bushes and the family moved into the water and out to the centre of the swamp. In the six and a half hours between arriving at the swamp in the morning and incident with the Ravens, two ducklings had disappeared. On average, 20% of the ducklings fail to survive to two months old, at which stage the birds are able to fly. This means that the island produces a surplus of birds which must leave in order to eventually establish territories for themselves.

As the ducklings reach adult size the family groups break up as they begin to congregate in November and into early summer. It is at this stage the peak population for the year is reached. The minimum population is reached in the laying season and in the early 1980s the lowest population recorded was 79 birds in July 1982. The annual cycle then begins again as numbers build up as a result of young of the year entering the population.

The Mountain Duck feeds on vegetable matter grazed on or around the lakes and swamps as well as eating invertebrates such as chironomid larvae, ephydrid larvae and pupae as well as ostracods. They have been known to eat adults and larvae of dragonflies, beetles and midges.

GREY PLOVER *Pluvialis squatarola* (Plate 10).

Identification: Length 27-31 cm; weight 140-380 g. This is a large plover with a big head and a solid, shortish bill. It is a visitor outside its breeding season and is most commonly seen in eclipse or non-breeding plumage. In this phase of its annual plumage cycle the upper surfaces are grey-brown with flecks of white, the crown is darker and the under surfaces are white, with the exception of the axillaries which are black. This gives the bird one of its most conspicuous features, a prominent black armpit which shows as it flies. The eyes are large and dark with the a white stripe above, the bill is black and the legs are dark grey. The tail and rump are white with brown-black bars on the tail.

The breeding plumage is predominantly black, hence another common name 'Black-bellied Plover'. As it occurs on Rottnest Island at the end of the breeding season or before departing at the beginning of the breeding season, it may have varying degrees of black on it in autumn or spring.

Status: The Grey Plover is a transequatorial migrant which breeds in the Arctic Circle, from the southern shores of the White Sea eastwards to the Canadian tundra during the northern summer (Fig. 4). As these areas are covered with the snow during the winter, the birds leave at the end of the breeding season and migrate to the southern hemisphere. No one is sure of the exact route but it probably travels over China before striking south for Australia.

It has been calculated that some migrating species of waders (or shore birds) may be able to fly up to 5000 km non-stop before pausing to feed and gain strength for the next leg. As it may take about a fortnight to gain this strength and four or five flight stages may be required for the migration north — south or vice versa, it probably takes the birds about two and a half months to make the journey. The Eighty Mile Beach (in North Western Australia) is a major staging area for this and many other waders which gather there on their migration.

The Grey Plover has been seen on the island every month of the year, but numbers do fluctuate on a seasonal pattern. Obviously, as they breed in the northern summer, numbers will be lowest on the island during our winter; the least we recorded during any visit was 2 in July 1983. Numbers start to build up between September and December (peak of 43 in February

1984) and drop in March and April as the birds start the long trek north.

This species may be seen on the beaches and exposed reefs all round the island and also commonly occurs on all of the lakes and occasionally on the swamps. The Grey Plover is often found with other wading birds but it is usually the only one of its species in a group of feeding waders. Nearly 80% of our 362 sightings of groups of Grey Plover have been of a lone bird, but occasionally they do occur in flocks; the largest we recorded was 20.

The Grey Plover eats crabs, sand shrimps, marine worms, beetles, weevils, hymenopterans (wasps and flies), caterpillars, amphipods and seeds.

RED-CAPPED DOTTEREL *Charadrius ruficapillus* (Plate 10).

Identification: Length 14-16 cm; weight 25-45 g. This attractive bird is one of the smallest wading birds found on the island. The same size as the Red-necked Stint, which is more common when present, they may be told apart by colouring and stance. The Red-capped Dotterel tends to stand much more erect and often bobs its head when stationary.

The male Red-capped Dotterel is pale brown on the upper surfaces, white on the lower surfaces, white on the forehead while the crown and nape are a lovely rufous colour. A black stripe divides the white forehead from the rufous cap and a black bar extends from the base of the bill through each eye. The bill is short and black and the legs are black. The females do not have the rufous cap or the black line over the forehead.

Status: The Red-capped Dotterel is a common resident. Although there are always some birds present, numbers fluctuate seasonally, with the population being highest in summer (peak recorded was 184 in February 1983) and lowest in winter and early spring (4 in June 1984). It occurs most commonly around all of the salt-lakes but has been seen occasionally on Salmon and Bickley Swamps. This species is seen less often on the beaches. We have only seen them there on 9 separate occasions out of nearly 1000 sightings. They are usually seen either singly or in pairs with just over 70% of our 963 sightings being of this nature. Flocks of 21 or larger were seen about 2% of the time and the largest flock we recorded was about 150 birds on the runway verge beside Government House Lake.

The Red-capped Dotterel is one of the few species of wading birds to breed on the island. They nest in a scrape on the ground which may be lined with bits of shell or dried leaves of plants. The nests may be around the edge of the lakes, in the limestone quarry towards West End near the turnoff to Wilson Bay, or occasionally above the high water mark on beaches. The breeding season is extended as the two eggs may be laid any time between June and March. The adults use a 'broken wing' display to distract any intruders approaching the nest and the cryptic colouring of the eggs makes them hard to find.

This species often forages with other species of wading birds and it feeds on insect larvae, ants, beetles, worms and possibly seeds.

RED-NECKED AVOCET *Recurvirostra novaehollandiae* (Plate 9).

Identification: Length 42-48 cm; weight 190-300 g. The Red-necked Avocet has a chestnut head and neck, and a long thin upturned bill. These two features are diagnostic and distinguish it from the Banded Stilt which is a similar size and shape. The body of the Avocet is white and the wings are brown-black with a broad white band through them. The legs are very long and pale grey-black, the toes are webbed and the bill is black.

Status: During the late 1950s and early 1960s the Red-necked Avocet was regarded as a rare visitor. The status of this species has changed since then. It is now resident, but in small numbers, breeding in several locations on the island. Despite this small population, their numbers do fluctuate seasonally, being lowest in late winter and early spring (less than 5 birds) and highest in summer. They usually occur in small groups of less than 5 but flocks which may contain all of the birds on the island may be seen, as in December 1981 when 26 birds were feeding on Pink Lake.

The Avocets have been seen on Government House, Pink, Garden, Herschell, Baghdad, Serpentine, Timperley, Negri and Vincent Lakes, and Salmon, Bulldozer and Parakeet Swamps. Of these, they appear to favour Pink, West Baghdad and Government House Lakes.

Red-necked Avocets have nested on the exposed substrate on Salmon Swamp, and possible on the edge of West Baghdad and Serpentine Lakes. The nest is a scrape in the bare ground and it may have a few traces of dried vegetation as a lining. The 4 eggs are laid between November and January but little is known about breeding behaviour. The parents lead intruders away with 'broken wing' displays and fly at intruders 'tooting' loudly if they approach the nest or chicks. The chicks leave the nest soon after hatching, following the parents around. At the first sign of danger they scurry for cover in a rock crevice or grass tussock and 'freeze', relying on their cryptic colouring for protection.

Silver Gulls and Ravens are predators of young Avocets and the adults do not like them near their chicks. We watched two adults and 3 small downy chicks foraging on the edge of Baghdad Lake when 3 Silver Gulls swooped down on the chicks. One adult Avocet took off and continually flew at the Gulls chasing them well away from the area. While this was going on, the other adult led the chicks into the water and out into the centre of the lake. Unfortunately the adults are not always successful in chasing off predatory Gulls or Ravens as we saw a Silver Gull kill an Avocet chick which had become separated from 2 adults and 5 other chicks. Several immature birds banded in summer were still present on the island six months later.

The long upturned bill is used to sweep through the water in search of food which consists of Brine Shrimps, aquatic insects and vegetable matter.

BANDED STILT *Cladorhynchus leucocephalus* (Plate 9)

Identification: Length 35-43 cm; weight 160-260 g. The Banded Stilt has a snow white head and body, black wings (with the exception of the trailing edge which is white and shows as a wide terminal bar in flight), a long finely tapered black bill and long pink legs. The adults have a broad chestnut band across the chest extending down to the abdomen. This is lacking in birds in their first year, then progressively develops with age.

Status: The Banded Stilt is a common non-breeding visitor but the number visiting fluctuates widely, both between and within years. The basic pattern of occurrence is for the largest numbers to be present in summer and the lowest in winter, when they may be entirely absent. This difference between years is shown by population counts made between December 1981 and October 1984. Peak recordings were of 847 birds in February 1982, 3192 birds in March 1983 and 1004 in February 1984. There was a four-fold difference between the peak numbers in 1982 and those of the following year (Fig. 5a). During this survey period the lowest recordings were of 4 birds. It is possible that these were birds which were unable to fly because of wing damage or some other physical abnormality. This pattern of occurrence appears to have been little changed as long as records have been maintained, although the exact timing of the birds' arrival and departure differs between years.

The birds feed on Brine Shrimps of which there are an abundance in the salt-lakes. This is one of the few species on the lakes, with the exception of the ducks and the Silver Gull, which feeds while swimming and food is obviously so plentiful that the birds are able to spend long periods resting on sandbars in the salt-lakes. After a bout of feeding, the birds fly to join others on the edge of the lakes. They crowd in and form what appears from the distance to be tightly packed 'rafts'. If startled, these congregations take to the air, almost as a single unit, wheeling in a well co-ordinated flock, coming back to land when the disturbance has ceased. Government House and Herschell Lakes are the areas on which the birds are most commonly seen and the 'rafts' of resting birds may be seen on the sandbar to the west of Vlamingh's Lookout (on Government House Lake) or on the bar stretching from the west end of the islet in Herschell Lake. Both of these sites can be seen from the Causeway.

The birds have been seen on all of the other salt-lakes as well as Lighthouse,Salmon and Bulldozer Swamps, where they are only found in small numbers.

The Banded Stilt is also known as the 'Rottnest Snipe' and during the last century was shot in large numbers for sport. This was done from brushwood brakes around the lakes with prisoners being employed to beat the birds towards the shooters.

One of the fascinating things about this species is that its nest and eggs were not described until 1930, 114 years after the bird itself was discovered by Europeans. Since then there have been only 19 records of breeding for this species. They breed on the inland salt-lakes of the arid and semi-arid areas of Austrlaia's southern interior. As they feed on Brine Shrimps, whose eggs are activated and hatched only in water, the birds may only nest when these lakes have water in them. This is an irregular event, but when breeding does take place, it does so in enormous numbers. Andrew Burbidge and Phil Fuller of the Western Australian Wildlife Research Centre found a recently abandoned colony on islands in Lake Barlee in September 1980. They estimated that there were 179,000 nests in the area and these would have produced over half a million eggs. From their survey they calculated that about 255,000 addled eggs and dead chicks were scattered round the colony. Despite this mortality, about 250,000 young birds probably survived to leave the colony. Even allowing for further heavy mortality among these young birds, the increase to the total population of this species would have been significant.

As the birds are dependent on flooding of the inland salt-lakes for breeding to take place, it is not surprising that breeding is irregular and that the birds are sensitive to cyclonic rains. In the middle of March 1983 there were widespread and heavy rains through the South-west, Goldfields and out into the Eucla. As these rains were finishing the birds on Rottnest Island started to leave for the mainland. On the 20 March there were just over 3000 on the island and 24 hours later there were less than 200 present. This phenomenon has been noted before as Glen Storr recorded that in 1955 the birds left the island in the latter part of February after unseasonal heavy rains. The birds did not return until the next November and breeding had taken place in the interim. This was shown by the fact that some of the incoming birds lacked the chestnut chest band of the adults, indicating they were birds in their first year of life.

RUDDY TURNSTONE *Arenaria interpres* (Plate 10)

Identification: Length 21-24 cm; weight 100-170 g. The short orange legs, the distinctive plumage pattern, the tubby shape and the foraging habit of turning stones to look for food underneath, makes this bird very easy to identify. The upper surfaces are dark brown, the throat is white, the chest has a broad brown-black band across it, the lower back is white (which only shows as the bird flies), the lower surfaces are white and the bill is black. This species may be seen in partial or full breeding plumage, even though the birds do not breed in the Southern Hemisphere. In this plumage phase the upper surfaces have chestnut and black mottling and the band on the chest and the brown around the face is replaced by a distinctive contrasting black and white pattern which is similar to that of the Magpie-lark or Peewee *Grallina cyanoleuca*.

Status: The Ruddy Turnstone is a common non-breeding visitor which breeds in the high Arctic tundra during the Northern Summer and spends the non-breeding season in the Southern Hemisphere. Their migration routes probably follow a similar pattern and timing to those of the Grey Plover which makes the same journey (Fig. 4). The bird is seasonal on the island, with largest numbers in the summer (peak of 495 in December 1983) and lowest in the winter (17 in July 1982). There are usually Turnstones on the island, even during their breeding season. These are probably immature birds which overwinter in Australia while the adults return to Siberia to breed. The bulk of the population arrives in September and begins to leave in April/May. It is during these transition periods when birds in partial or full breeding plumage may be seen. Numbers vary over the summer as birds move to and from other suitable foraging areas in the south west.

The Ruddy Turnstone seems to favour shores which are covered with stones or weed. These are ideal sites for its foraging technique which consists of flipping over stones and debris with its robust bill and seizing prey sheltering underneath. They can move quite large stones and we saw a Turnstone tip up a stone weighing 267 g which may have been nearly twice as heavy as the bird itself. They also feed by burrowing into the edges of weed banks or sand covered patches of beach washed sea-grass. Their diet consists of amphipods, beetles, flies and their larvae, slaters, spiders and other invertebrates as well as vegetable matter. We have seen them feeding on bread crumbs which they pulled from a loaf of bread lying on the beach at Geordie Bay.

The edges of the salt-lakes are the preferred feeding areas and 85% of the 8000 plus birds we recorded on the island were on these sites. Of the lakes, Government House, Herschell and Serpentine Lakes appear to be the favourite sites with nearly 80% of all of our observations being made there. Foraging also takes place round Baghdad, Timperley, Vincent, Pink and occasionally round Garden and Sirius Lakes. They have also been seen around Parakeet and Bickley Swamps. Virtually all of the beaches and exposed reefs round the island and its associated islets and stacks are visited by foraging Turnstones and we have twice recorded birds roosting on the sandblow at West End beside the turnoff to Radar Reef.

Rottnest Island is an important conservation area for this species as it has a large expanse of suitable foraging sites, unlike the adjacent mainland, where such opportunities are limited. The importance of the island is illustrated by the following figures. In 1983 the Royal Australiasian Ornithologists Union estimated that there were 5347 Ruddy Turnstones in Australia. If that figure is even remotely accurate, nearly 9% of the Australian population was concentrated on the 200 ha of salt-lakes on Rottnest Island.

GREY-TAILED TATTLER *Tringa brevipes* (Plate 10).

Identification: Length 24-29 cm; weight 70-150 g. The upper surfaces are plain grey, there is a white eyebrow which meets above the bill and the body is very sleek. The breast is pale grey and the under surfaces are white. The bill is long, straight and grey and sometimes has a green-yellow coloured base. The legs are a moderate length and green-yellow in colour.
Status: The Grey-tailed Tattler is a transequatorial migrant which breeds in north-eastern Siberia and spends the non-breeding season in the Southern Hemisphere. The migration route is probably similar to that of the Grey Plover described elsewhere in this book (Fig. 4).

This species, like many other transequatorial migrants, is more common in the north of the State than the south-west; nevertheless, it is a regular visitor in small numbers (less than 20) to Rottnest Island and Grey-tailed Tattlers may be found here all year. The birds present over winter are probably immature birds which have remained while the adults return to Siberia to breed. On the island it forages around the salt-lakes or the exposed reefs round the coast. Of the lakes, Serpentine and Herschell appear to be favoured as feeding areas, with Government House and Baghdad accounting for about half of the total sightings of the other two lakes. Tattlers have been seen foraging on most of the exposed reefs with Wilson Bay, Cape Vlamingh and Mabel Cove being the three areas most commonly visited. They are usually seen alone and even when feeding with other species of wading bird, there is usually only one Tattler in the flock. They do occur in groups, usually of two birds, and 10 is the largest group seen on the island.

Little is known of their diet, but they have been reported as eating insects (including mosquitoes) and caterpillars.

RED-NECKED STINT *Calidris ruficollis* (Plate 10).

Identification: Length 14-17 cm; weight 22-42 g. The Red-necked Stint is the most common and the smallest sandpiper occurring on the island. It is grey-brown on the upper surfaces, with darker streaking and mottling, it has a slight dark grey-brown line from the base of the bill through each eye, with a white line above. The grey-brown of the back is continuous around the side of the neck and becomes streaky on the upper breast. The under surfaces are white, the eyes are dark brown, the bill is short and black, as are the legs.

At certain times of year, before leaving for or after returning from the breeding grounds, some birds may be in partial breeding plumage and it is apparent then how it received its common name. This plumage phase is characterized by a deep salmon pink or russet brown colouring on the head, neck and upper breast with chestnut mottling on the remaining upper surfaces.

Status: This species is a transequatorial migrant which breeds in eastern Siberia and western Alaska and during the non-breeding season migrates to Australia as well as countries in South-east Asia and New Zealand (Fig. 4). It is fascinating that an animal weighing around 30 g will undertake this long migration twice a year, all of its adult life. Recoveries of birds banded in Australia indicate that they travel to and from their breeding grounds via Indonesia, the South China Sea, Vietnam, China, Japan and probably up the east coast of the U.S.S.R.; a round trip of about 26,000 km per year.

The Red-necked Stint is the most abundant transequatorial migrant wader occurring in Australia. It is believed that around 300,000 spend their non-breeding season in Australia, most of them in Southern Australia. The number of birds on Rottnest Island fluctuates with season, being lowest in winter when breeding is taking place in the Northern Hemisphere, and highest in summer. There are usually some birds present all year although the number remaining varies from year to year. For example, over 200 remained during the winter of 1982, while only 30 remained in 1984 (Fig. 5b). Those present are probably immature birds not ready to breed or birds physiologically incapable of leaving the island. The migrating birds begin to arrive in late August and early September and some of these may be in partial breeding plumage. It is intriguing to see birds feeding together, which because of their different colouring, appear to be different species. During this build-up phase flocks of several hundred birds arrive and join those already there. The maximum number is reached in early December and peak numbers vary from year to year. For example, there were over 3500 birds present in early December 1982 but just over 1600 at the same time in 1983. The number of birds present on the island fluctuates throughout the summer and autumn. These fluctuations are the result of both local movement around the Swan Coastal Plain and of birds staging through the island on their way further south in spring and on their way north in autumn. The birds begin their migration north in April or May with numbers falling rapidly over this period.

On the island the birds forage in flocks of up to 450, mixed with other species of wader. They feed along the margins of the salt-lakes and swamps and, rarely and in small numbers, around the coast. The bulk of the population feeds around Government House and Baghdad Lakes. In fact over 75% of our 35,000 plus sightings of stints have been on these two lakes. Two of the most convenient observaion areas are the Causeway between Government House and Herschell Lakes or the road near the turnoff to Little Parakeet Bay, which overlooks a freshwater soak on Baghdad Lake, where large flocks of mixed waders often forage. The Red-necked Stint feeds on mosquitoes, weevils, hymenopterans, insect larvae, beetles and, possibly seeds.

CURLEW SANDPIPER *Calidris ferruginea* (Plate 10).

Identification: Length 20-25 cm; weight 45-88 g. The Curlew Sandpiper has grey-brown upper surfaces, a white stripe from the base of the bill, through and extending behind each eye, white under surfaces, long black legs and a long thin distinctly down-curved black bill. In flight this species may be distinguished by a white stripe along the wing and its white rump.
Status: The Curlew Sandpiper is a transequatorial migrant which breeds in northern and eastern Siberia in large numbers and winters in many parts of the Southern Hemisphere, including South-east Asia, Africa, Australia and New Zealand (Fig. 4). This species is the next most common wader to visit Australia after the Red-necked Stint. It is a common non-breeding visitor to Rottnest Island with numbers fluctuating seasonally, showing the typical pattern of such migrants; numbers being lowest in winter and highest in summer. Peak numbers vary between years; for example, there were just over 400 present in late November 1982, but over 700 at the same time the following year. The bulk of the birds leave for the breeding areas in April and May with the lowest numbers being present over winter (3 birds in 1982; 4 in 1983). The total number of birds on the island changes constantly as some birds move locally to various suitable sites around the Swan Coastal Plain, as well as others staging through the island as part of their migration route to locations further south.

They are usually found in mixed flocks with other waders, usually the Red-necked Stint and Ruddy Turnstone. With its longer legs and bill, the Curlew Sandpiper is able to feed in deeper water than the other waders, probing into the lake floor for food. It is known to feed on insects (both adults and larvae), worms, crustaceans, molluscs, bugs, as well as some vegetable matter. They frequent the margins of the salt-lakes and swamps and, very occasionally, the coast. Of the salt-lakes, Government House and Baghdad Lakes are favoured and about 60% of over 6000 observations of Curlew Sandpipers have been made round those lakes. As with the Red-necked Stint, the Causeway and the road near Parakeet Soak on Baghdad Lake are the best vantage points.

CASPIAN TERN *Hydroprogne caspia* (Plate 11).

Identification: Length 49-60 cm; weight 540-680 g. The Caspian Tern is the largest of the four species of tern which breed on the island. Its size and distinctive large, bright red bill make it unmistakeable. The upper surfaces are grey, the crown and part of the back of the neck black, the under surfaces white, the eyes dark brown and the legs black. When adults are not in breeding plumage the black crown becomes streaked with white as it is in immature birds.

Status: The Caspian Tern is a conspicuous resident which is usually seen singly or in pairs. The birds may be seen around most of the coast as they roost on beaches or fly over the water hunting for fish which they catch by diving. They are often found around the salt-lakes where they roost on the lake margins or wash in the fresh water soaks. The shell beach on the north eastern corner of Government House Lake beside the road to the airport is one such roost site where this species may often be seen, sometimes in flocks. We have seen groups of up to 38 birds at this spot and this may be the total population of the island.

This species has increased its numbers since the late 1950s when it was regarded as an uncommon resident, with single pairs breeding on offshore islands and stacks like Green, Dyer and Parakeet Islands and Clune, Armstrong and Crayfish Rocks. There is now a small breeding colony which nests on the islet in Herschell Lake. In August 1983 there were 7 pairs breeding in one area, about 3 m wide and 10 m long, between a Crested Tern colony to the west and Silver Gulls to the north and east. The Caspian Tern nests in a scrape in the ground and lays 1-2 eggs some time between the beginning of July and the end of October. Little is known of the breeding behaviour, but incubation takes about 3 weeks and the chicks must stay around the nesting area until they are able to fly, at which stage the parents move away from the nesting area. Adults may be seen flying in from the coast, carrying fish for the young.

This species is very aggressive, readily flying at anyone it regards as being too close to its nest or its free flying young. They will chase other birds as well as humans and we have seen them leaving their nests on Herschell Islet to fly, calling loudly, at people standing on the Causeway, over 200 m from the colony.

FAIRY TERN *Sterna nereis* (Plate 11).

Identification: Length 20-27 cm; weight 57-85 g. This is the smallest of the Australian terns and the upper surfaces are very pale grey, the black on the top of the head extends down the back of the neck, the forehead is white, the under surfaces are white, the bill is yellow, the eyes are brown and the legs orange.

Status: The Fairy Tern is a common breeding visitor arriving in spring, breeding over summer and leaving in autumn. The timing of their arrival varies from year to year and it may be any time from early September into October. The number of birds present builds up to a peak in December or January and they begin to leave in February or March. Some individuals may remain around the island until early winter as we have seen several birds around in early June during both 1982 and 1983. Little is known about the movements of the birds once they leave the island, but it is believed that they move north to the West Kimberley coast. When present they are noisy and conspicuous, flying to and from their breeding colonies, keeping up a constant 'chittering' as they forage. They may often be seen in small groups flying close to the shore as they search for fish which they catch by plunging into the sea; or roosting in small numbers on beaches such as the east end of Salmon Bay or at Phillip Point. They are also vocal and aggressive if disturbed when breeding and they will fly very close to any intruder and they readily fly over and spatter unwelcome visitors with droppings.

The Fairy Tern lays 1-2 or occasionally 3 eggs in a scrape in the ground on shell beaches around the salt-lakes or the coast. The location and number of these breeding colonies may vary from year to year. The birds are very 'flighty' and easily disturbed, with colonies being deserted as a result of some disturbances. There is some concern at present about the future of this small tern as a result of changes to some of its breeding habitat around Australia. A brief history of the breeding of this species on Rottnest Island will serve to explain some of the problems being faced by it.

These lovely little terns have been recorded as breeding visitors to the island since bird watchers first commented at the turn of this century. Dom Serventy found eggs and chicks on a small islet in one of the salt-lakes in January 1925 and in the late 1920s during December and January, one visitor found a colony breeding on the shore of one of the lakes. He noted that the colony contained about 200 nests but most of these were destroyed by vandals, something which still happens occasionally today.

Vincent Serventy wrote an account of the breeding of Fairy Terns during the summers of 1946/47, 1948/49 and 1949/50. In early February 1947 he found a colony on the spit jutting into the north shore of Government House Lake (probably the spit to the west of Vlamingh's Lookout) but he gave no indication of the size of this colony. A month earlier his brother Dom found a colony of about 40 nests near the old salt works on the west side of Government House Lake. This colony was destroyed by vandals.

In late January 1949 Vincent Serventy found a colony on the sand bar which separates Baghdad from Vincent Lake. There were 60-69 nests in this colony and these were at various stages from eggs to nestlings. Serventy commented that the area was often visited by holiday makers and that there was the possibility of eggs being destroyed by accident.

In late January 1950 Vincent Serventy found a colony of 234 nests on the small islet in the west section of Baghdad Lake. This colony ranged from eggs to large nestlings, was the largest colony seen by Serventy and it was also the most well protected as it was surrounded by water and the lake floor was (and still is) like 'quick-sand' and very difficult to move across. It is obvious from his account that the Fairy Tern colony occupied the whole island and that it was a reasonably successful colony. It was apparent that breeding numbers then were much the same as they were in the 1920s.

Glen Storr also produced an account of the breeding of this species. During December 1956 colonies were established on beaches in Strickland Bay and Cape Vlamingh, but high tides caused these attempts to fail. In November 1957 a successful colony was established along the north side of the western edge of the airstrip. The following November the birds nested on the runway itself but most of the eggs and many of the adults were destroyed by aeroplanes.

In December 1975 Ian Abbott and colleagues at the University of Western Australia found a colony of about 50 nests near Cape Vlamingh adjacent to the road. This colony failed. In December 1977 about 100 birds attempted to breed there but human disturbance probably caused the birds to leave their nests, allowing Silver Gulls to eat the eggs.

There was a large colony on the sandbar between Baghdad and Vincent Lakes in February 1980 (the site of the January 1949 colony). Breeding was well advanced as there were a number of large nestlings, capable of flight, being fed by adults. The remainder of the colony varied from eggs still being incubated through to chicks about to leave the nest. This colony contained between 200 and 250 nests. The following summer they nested at the west end of the islet on Herschell Lake. Around Christmas 1980 there were about 190 nests and breeding was probably at an early stage as 30 nests were checked (by telescope from out in the lake) and they contained either one (25) or two (5) eggs. No nestlings were seen. This site was used the following summer. In late December 1981 there were 22 nests and by early February 1982 there were 110 varying in stage from eggs to chicks.

In the summer of 1982/83 there were two colonies: one on a spit which partially separates the west section of Baghdad Lake from the east section; the other was on the spit jutting into Government House Lake to the west of Vlamingh's Lookout. The former colony contained about 50 nests in early December 1982 and the latter about 130. Nothing is known about the success of the latter colony but the former appeared to be successful.

During the summer of 1983/84 there were at least four colonies scattered over the length of the island. The smallest were each of about 6 nests. One was on the sand blow at West End beside the turnoff to Radar Reef and the other between the runway access road and the south side of Government House Lake. Another colony of over 120 nests was on the islet in Baghdad lake during January 1984 and the largest colony was on the shell beach at Cape Vlamingh. In mid-December 1983 it contained about 200 nests ranging from eggs to large young about to fly. This colony put up with considerable disturbance from people walking through it while fossicking for shells or fishing. Despite this disturbance it was apparently successful in producing young.

The following summer (1984/85) there were again at least 4 colonies over the island. All were occupied in late December 1984 and they consisted of: 30-50 nests on the shell beach at Cape Vlamingh; about 60 nests on the spit partially dividing east from west Baghdad Lake; about 20-25 nests on Ricey Beach and ; about 100 nests on the south side of Government House Lake between the stone bank and the runway verge. The Rottnest Island Board bus drivers said that the colony at Cape Vlamingh was being constantly disturbed and we saw people walking through the colony and lying on the beach at the edge of the colony. This kept the adults off their nests and Silver Gulls and Ravens were hanging around the colony looking for easy pickings. The colony beside the runway was also constantly disrupted as planes and buses caused the birds to take to the air. The buses were more disruptive as they ran right along the edge of the colony.

It appears that the number of birds breeding on Rottnest Island has remained at about 200-300 pairs for the last 60 years and that these birds constantly change their breeding sites, some of which are prone to failure through continued disturbance. Anything which results in the adults being off the nest for long periods will allow Gulls and Ravens to steal their eggs and such risks should be minimized. In order to manage Fairy Terns to ensure their breeding success the Rottnest Island Board should arrange to temporarily close off sections of beaches or shores of lakes to allow colonies (once established) a chance to be successful with a minimum of disturbance. Such closures need only be brief and they may well help the Fairy Tern to continue to breed successfully on the island.

WHITE-FRONTED CHAT *Epthianura albifrons* (Plate 13).

Identification: Length 11-13 cm; weight 11-13 g. A distinctive small bird often seen perched at the top of low vegetation around the lakes and swamps, or on the power lines along the centre road. The male is the more distinctive being predominantly black and white. The face, forehead, breast and throat are white. This area is enclosed by a broad band of black which covers the crown, back of the head and much of the chest. The back is grey, the tail black, the eyes brown, the legs and beak are black. The female and immatures lack the black on the crown and back of the head and their chest band is brown-black.

Status: The White-fronted Chat is a common resident. It may be seen in small parties of up to 8 birds, but it does form larger flocks of varying sizes up to 80-90 birds which may occur at any time of year.

It is found around all of the salt-lakes and swamps where it forages in the low vegetation looking for insects which form the bulk of its diet. It also forages on the bare shores of the lakes where its feeding technique resembles that of the Ruddy Turnstone, turning over pebbles and other debris looking for insects sheltering underneath. We once saw a White-fronted Chat turn over a stone weighing 11.7 g to catch something lurking underneath. On another occasion we saw a male act like a bulldozer and push aside a piece of wood weighing 22.7 g. This piece of wood may have been twice as heavy as the bird itself. The Chats also forage through the heath and on the golf course. They may be seen the length and breadth of the island, including the coastal beaches where they forage almost down to the water line.

This species has an extended breeding season and egg laying may take place any time from the beginning of June until the end of December. Kent Williams studied the White-fronted Chat for 4 breeding seasons on the island during the late 1960s. He found that egg laying is dependent on winter rainfall and in years with reduced rainfall there is a corresponding reduction in nesting effort. With the start of breeding, pairs establish a territory within which the female builds a cup-shaped nest of twigs and fine grass, etc. Nests are usually built in low vegetation, within 1 m of the ground and favoured areas are the margins of the lakes and swamps, but they may build in the heath away from these areas. The male defends the territory chasing other chats from the area. Both adults will attempt to lead intruders away from the nest, or fledged young by giving 'broken wing displays'. The 2-3 eggs are laid at daily intervals and both adults incubate the eggs which hatch about 2 weeks later. The nestlings are fed by both parents, leaving the nest as free-flying individuals 2 weeks after hatching. About one third of all eggs laid give rise to free-flying young. The family leaves its territory and joins with other birds to form foraging flocks. No one knows how long the parents maintain their young but the young may leave the island as one bird banded as an immature was recovered on the adjacent mainland 9 months after banding.

The White-fronted Chat may begin to breed in its second year of life and Williams found that 5 of the 100 birds he banded as juveniles or adults lived for at least 3 years, but some individuals may live much longer than that.

SECTION III: Birds of the Swamps

Black Duck

BLACK DUCK *Anas superciliosa* (Plate 8).

Identification: Length 47-62 cm; weight 0.8-1.4 kg. This duck is readily identified by its characteristic white-striped pattern on the head. The crown is black, there is a thin white stripe above each eye, a black stripe through the eyes and a broad white stripe from the base of the bill extending under the eyes to the back of the head. The throat and lower half of the face is dirty white, the upper surfaces of the body are black-brown and most feathers are edged with buff. The under sides of the body and tail are brown-black while most of the under sides of the wings are white. There is a rectangular area of green on the wing called the speculum, the eyes are brown, the bill is grey and the legs are dull yellow.
Status: The Black Duck is a migrant from the mainland which does not breed on the island. It has been recorded at all times of the year with the exception of June or July. It is usually seen in small groups of about 5 birds; the largest flock seen to date has been 20. The population on the island at any time is usually small as 24 is the largest number recorded and these were in two flocks of 15 and 9 in March 1984. This duck is often seen in association with the Grey Teal; Salmon and Lighthouse Swamps being its favourite haunts. It has been recorded occasionally from Bickley Swamp, Lake Baghdad, Pink Lake, Parakeet Swamp and Lake Negri.

The Black Duck feeds by dabbling its feet in shallow water, stripping seeds from plants and by up-ending itself in deeper water. Its diet consists of seeds of swamp plants and insects (water beetles and water boatmen).

GREY TEAL *Anas gibberifrons* (Plate 8).

Identification: Length 37-48 cm; weight 350-670 g, with the male being slightly larger than the female. This duck is about two-thirds the size of the Black Duck with which it often associates. The upper surfaces are grey-brown, the feathers have a pale margin and the white throat is a characteristic feature. The speculum is black with green, the cheeks are pale grey while the crown and the back of the head are black-brown. The underside of the body is pale brown, the eyes bright red-brown, and the bill and legs are black.
Status: The Grey Teal is a regular non-breeding visitor occurring in small numbers. The average size of the flocks seen is 12 and groups have ranged from 3 to 24 birds. The largest number of birds recorded was 25 in July 1982 and like Black Duck, they favour Salmon and Lighthouse Swamps. They have been seen once at Bickley Swamp and once beside Government House Lake.

The Grey Teal feeds by dabbling at the water's edge, stripping seed from plants and by up-ending in the deeper water. It feeds on seeds of swamp-vegetation and aquatic insects including larvae of midges, dragonflies, water beetles and water boatmen.

Banding studies elsewhere in Australia have shown that Grey Teal may live in excess of 21 years.

Plate 1. Coastal habitats

1a. Looking west from North Point, showing rocky headlands, reef platforms and islet (Armstrong Rock).

1b. Fish Hook Bay showing cliffs and exposed reef platforms.

2a. Wallace Island (from Bickley Point). Note exposed reef platform and the two Osprey on a stick nest at the highest point on the islet.

2b. Green Island from the sandy beach at Nancy Cove.

Plate 3. Salt-lake habitats.

3a. Herschell Lake (looking north from the Causeway). Note the expansive exposed beach and the limestone rubble at the bottom right of the photograph. This is the edge of the road to the settlements at Geordie and Longreach Bays.

3b. Sirius Lake with high water levels. This lake is ephemeral, being dry in later summer. During winter the water reaches the fringing samphire. Note the Pied Oystercatcher.

Plate 4. Woodland habitats.

4a. Serpentine Lake from the air. Note the beach and just beyond, the remnant *Melaleuca* woodland. In the background is an area of plantation surrounded by heath. The line across the upper quarter of the photograph is the remains of the old railway line between Kingstown Barracks and the guns on Oliver Hill.

4b. Remnant *Melaleuca* woodland fringing the north side of Serpentine Lake.

Plate 5. Woodland and heath habitats.

5a. View from Wadgemup Hill looking north-east over Baghdad Lake to the settlement at Geordie Bay. Note the *Acacia* woodland in the foreground, the plantation in the right half of the photograph and the extensive heath.

5b. Plantation west of Wadgemup Hill, surrounded by heath dominated by prickly *Acanthocarpus preissii.*

Plate 6. Heath and disturbed areas.

6a. Coastal heath in the foreground, Crayfish Rock in the background.

6b. The runway verge near the turnoff to the aircraft park. This artificially short vegetation favours the Banded Plover.

Plate 7

Little Pied Cormorant
Wedge-tailed Shearwater
Pied Cormorant
White-faced Heron
Reef Heron
♂
♀
Peafowl

Plate 8

Grey Teal
Black Swan
Black Duck
♂
♀
Mountain Duck
Nankeen Kestrel
♀
♂
Osprey
♂
♀
Ring-necked Pheasant

Plate 9

Pied Oystercatcher
Bar-tailed Godwit
Whimbrel
Banded Stilt
Red-necked Avocet

Plate 10

♂
♀
Red-capped Dotterel
Banded Plover
Grey Plover
Ruddy Turnstone
Curlew Sandpiper
Grey-tailed Tattler
Sanderling
Red-necked Stint

Plate 11

Silver Gull
Bridled Tern
Caspian Tern
Crested Tern
Fairy Tern

Plate 12

Spotted Turtledove
Laughing Turtledove
Rainbow Bee-eater
Tree Martin
Sacred Kingfisher
Welcome Swallow
Rock Parrot
Fan-tailed Cuckoo

Plate 13

Singing Honeyeater
Western Warbler
Richard's Pipit
White-fronted Chat
Spotted Scrub-wren
Silvereye
Golden Whistler
Red-capped Robin

Black-faced Cuckoo-shrike

Raven

SECTION IV: Birds of the Woodlands

Red-capped Robin

SPOTTED TURTLEDOVE *Streptopelia chinensis* (Plate 12).

Identification: Length 28-32 cm; weight 130-172 g. This dove is brown with mottled dark brown on the wings and upper tail. The under surfaces are pale pinkish-brown, the eyes are yellow, the bill grey and the legs pink. The most distinctive feature of this species (which distinguishes it from the smaller Laughing Turtledove) is the prominent black band with white spots on the back of the neck.

Status: The Spotted Turtledove is a native of China and Indo-Malaya which was introduced deliberately to Western Australia. The population in Perth descends from a group of birds brought from Melbourne to the South Perth Zoo where they were bred and released in 1898 or 1899. The birds gradually spread and the island population is a result of colonists which arrived around 1937. By the 1950s it was regarded as a moderately common resident with the bulk of the population occurring in the eastern half of the island, which was covered by more woodland in those days. The big fire of 1955 destroyed large areas of woodland and left isolated clumps of trees. This may have lead to a decrease in numbers of the Spotted Turtledove in the area between the west end of the salt-lakes and Narrowneck. Nowadays the birds may be seen in small groups of up to 3 birds throughout the wooded areas, east of Narrowneck, but most of the population is concentrated in the areas to the east of Herschell and Government House Lakes.

The Spotted Turtledove is mainly dependent on man for food, stealing food put out for poultry and collecting scraps round picnic and camping grounds. Rottnest Island is an ideal locality for this species as there are ample scraps from tourists who will readily feed animals waiting round picnic tables.

These birds have an extended breeding season but most nesting takes place in spring and summer. A rather flimsy platform of twigs is built in a tree and 2 eggs are laid. These are incubated for about 2 weeks and the nestlings remain in the nest for 2 weeks. They then join their parents foraging for food.

The Spotted Turtledove is much less common than the Laughing Turtledove which appears to outnumber it by 4 to 1.

LAUGHING TURTLEDOVE *Streptopelia senegalensis* (Plate 12).

Identification: Length 24-27 cm; weight 73-132 g. The Laughing Turtledove has brown upper surfaces, blue-grey wings and back, a buff band with black speckling around the throat, white tips to the outer tail feathers, brown eyes, a grey-brown bill and pink feet.

Status: Like the Spotted Turtledove, this species is an exotic, originating in East Africa. It was brought to Perth from Melbourne and released from the South Perth Zoo in 1898 or 1899. The Laughing Turtledove appeared to spread faster than the Spotted Turtledove and reached Rottnest Island about 1930; 7 years earlier than the larger Spotted Turtledove. By the 1950s the Laughing Turtledove occurred throughout the island but was more common on the eastern half. This is the situation today with this species being about 4 times more common than the Spotted Turtledove.

The Laughing Turtledove may be found in heath, along the margins of the lakes and in the woodland, although it appears to be more common in the wooded areas of the island particularly those sections east of Wadgemup Hill. It usually occurs in small parties but we have seen groups of up to 36 birds.

The diet consists of seed, food placed out for poultry and bread crumbs. The staff of the Rottnest Island Board sometimes feed the Mountain Ducks around Garden Lake and they keep bulk grain in a seed store which the Laughing Turtledove often visits in search of food. The birds obviously feed on seeds of vegetation occuring throughout the island so part of the population may not be dependent on humans and their scraps for food.

Breeding takes place over an extended period which may be affected by food supply, but most of the birds breed between September and March. The nest is a flimsy structure of twigs and it may be so poorly constructed that an observer standing underneath may see straight through the platform to determine the contents. The nest may be placed in trees at various heights but they also nest in buildings as we found them nesting in the sheds at the east side of Bickley Swamp. The birds lay 2 eggs which are incubated for just over 2 weeks. The nestlings remain in the nest for 2-3 weeks after which they join their parents foraging for food.

As both the Spotted and Laughing Turtledove are exotics, deliberately introduced to Perth over 80 years ago and reaching Rottnest Island under their own power, it must be asked if they should be removed from the island. This is a difficult question to answer, particularly as the only native pigeon (Brush Bronzewing *Phaps elegans*) believed to have occurred on the island became extinct at the turn of the century and so the Turtledoves are not competing with a native relative. The purists would argue, with some justification, that the birds are exotic, and therefore should be removed. This would prove difficult for two reasons. Firstly they are widespread on the adjacent mainland round Perth and would undoubtedly re-invade. Secondly, the island would need to be cleared up and policed to see that no food scraps are available to the doves (or the Raven and Silver Gull). Given that people like to feed animals (even exotics), this would prove very difficult on a resort like Rottnest Island and so the Turtledoves are undoubtedly with us to stay.

FAN-TAILED CUCKOO *Cuculus pyrrhophanus* (Plate 12).

Identification: Length 25-28 cm; weight 42-48 g. The adult's upper sur-

faces are slate-grey, under surfaces red-brown, long tail feathers dark grey and notched with white, eyes brown with yellow surrounds, bill black and feet yellow. The immatures are markedly different: the head and upper surfaces brown-black with rufous edges to the feathers, under surfaces are mottled pale grey and brown, and the tail is black with buff notches. The adult call is a trill which sounds like a pea being blown round in a tin whistle.
Status: The Fan-tailed Cuckoo is a regular migrant and it may be seen (or more commonly heard) in woodland areas, particularly the *Melaleuca* woodland and plantations. The birds arrive on the island in May or June and remain until October or November. The male sits on a perch, sometimes quite prominently displayed and produces a monotonous trill which may be heard at any time of the day or night.

Like many cuckoos, the Fan-tailed Cuckoo is parasitic, laying its eggs in the nests of other species, usually much smaller than itself. This cuckoo usually lays in dome-shaped nests and on Rottnest Island it parasitises Spotted Scrub-wren nests as an immature cuckoo was begging and being fed by Scrub-wrens in Bickley Woodland in late September 1982. The immature cuckoo, weighing nearly 40 g, was 4 times the weight of the host parents which would have had to work hard to feed a nestling so much larger than they were.

WELCOME SWALLOW *Hirundo neoxena* (Plate 12).

Identification: Length 15-16 cm; weight 10-20 g. The upper surfaces are blue-black, the forehead, throat and upper chest are chestnut, the under surface of the body is pale grey, the eyes dark brown, the bill is black, the legs brown and the tail is deeply forked with a white spot near the end of each tail feather. Immature birds lack the chestnut on the forehead and the throat and upper chest are duller than the adults.
Status: The Welcome Swallow is a common resident which may be seen in every habitat on the island, including the offshore stacks. Before Europeans provided buildings the birds were restricted to using caves as nest sites. These were only available around the coast of the main island and the offshore stacks. Such sites must have been fairly limited as they had to be protected and out of reach of the waves and heavy spray. These sites are still used today and the birds can be seen skimming the waves as they travel to and from their nest sites. With the arrival of Europeans the Swallows were able to nest on the island, rather than round it. It appears that as soon as the buildings were erected, the birds took advantage of them. The Army, with the provision of bunkers, lookouts and tunnels, provided many such sites. Most of the holiday cottages have resident Swallows which have a nest plastered somewhere round the building and the birds roost on the verandah rails, open doors, outside lights, etc. One tunnel, which was formerly an ammunition locker, has heavy pole supports for the ceiling and these poles provide ideal sites for nests. A colony of up to 25 pairs uses it at various times of the year.

The nest is built by both sexes and is a cup-shaped structure constructed of bits of vegetation cemented together with mud. This is anchored to the side of a vertical surface, with the birds taking advantage of an anchor, such as a nail, or some horizontal structure. The nest floor is lined with soft feathers collected from the ground (Gull, Mountain Duck, Raven, etc) or dried grass. The 2-4 (usually 3) eggs are laid sometime between August and November and incubated by both adults for just over 2 weeks. The chicks remain in the nest for between 2 and 3 weeks and remain dependent on their parents for a further 3 weeks. The adults are capable of rearing more than one brood in a season.

The same nest site may be occupied by the same pair which appear to remain together throughout the year. In the cooler weather the pair will roost on their nest, even when not breeding. In fact they appear to be so attached to the site that they will roost there, even if the nest has fallen down. This happened to a pair which had a nest attached to a nail on the eave of one of the buildings on Wadgemup Hill. For some reason the nest came unstuck and fell, leaving the nail and the bits of mud stuck to the eave. From then on the pair roosted side by side on the nail.

The Welcome Swallow is an insectivore, feeding on flies, midges and other insects which are caught in the air. The birds are common around the salt-lakes and swamps as they swoop low over the water, on occasion actually touching the water as they take an insect on the surface.

In the warmer weather the birds gather in large groups, sometimes of several hundred birds, and roost away from their nest sites. For example, the birds using the old ammunition locker roost in Tuarts on the hill above their nest site during the hot weather. The wires above the Bowling Black beside the Lodge provide another favourite roosting site and mixed flocks of several hundred Swallows and Tree Martins may often be seen there. As the wires run in the direction that the bowls are rolled, the line of excreta from these flocks must provide an interesting bias to the game.

There is an exchange between the Swallow populations on the island and the mainland and the surplus offspring may leave the island. One of the birds we banded was found dead at the Mobil Oil terminal at Leighton Beach and another was hit by a car along West Coast Highway, Sorrento Beach. Banding studies conducted elsewhere in Australia have shown that swallows may live for more than 7 years.

TREE MARTIN *Cecropis nigricans* (Plate 12).

Identification: Length 12-14 cm; weight 11.5-19.5 g. The upper surfaces are blue-black, except for a distinctive white rump which shows clearly in flight and distinguishes this species from the Welcome Swallow with which

it often mixes. The Tree Martin has a slight rufous patch on the forehead, the throat is grey, the other under surfaces are dirty white, the eyes are brown, the bill and legs are black and the tail is only slightly forked, unlike the Welcome Swallow which has a deeply forked tail.
Status: The Tree Martin, a common and regular visitor, is present in greatest numbers during the summer and early autumn. During the late 1950s and early 1960s this species was an uncommon visitor with a few birds (usually less than 5) visiting the island most summers. Now it visits in mobs of hundreds (600 plus) occurring in mixed flocks with the Welcome Swallow all over the island. Like the Swallow, the Tree Martin is an aerial insectivore; that is, catching its insect prey on the wing and the birds may be seen swooping low over the lakes catching prey, even on the surface of the water.

Although the majority of the birds are only visitors to the island and flocks have been seen crossing Cockburn Sound, small numbers appear to be present throughout the year. The Tree Martin nests in holes in trees but there are few, if any, such sites on the island, so the opportunities for nesting in traditional sites are limited. In August 1983 about 10 pairs did breed in a wooden tower on Wadgemup Hill. This building had a fully enclosed roof space with blocked off eaves, except for several small holes in the corners of the roof. The birds were collecting mud from a puddle near the Lighthouse Keeper's Cottage and carrying it several hundred metres to the tower and into the roof space. We could not inspect the nests but we banded several birds from this colony and all had brood patches and several had eggs which could be felt in their abdomens. This was the first breeding record for the Tree Martin on Rottnest Island and it is possible that this colony may remain, until the roof space is fully sealed off from the birds.

BLACK-FACED CUCKOO-SHRIKE *Coracina novaehollandiae* (Page 63).

Identification: Length 31-35 cm; weight 97-128 g. This lovely bird is steel (or battleship) grey, the forehead, side of the face and throat are black, as are the flight feathers on the wings and the tail is black with white tips. The under surfaces are white, the eyes, bill and legs are black. The immatures have very little black on the head; just a bar from the bill through and beyond the eyes and they have fine barring on the forehead and chest. The flight of the Black faced Cuckoo-shrike is characteristic as the birds flap and glide resulting in an undulating flight. When landing the birds settle their wings by flicking one then the other.
Status: The Black-faced Cuckoo-shrike is an uncommon, irregular, non-breeding visitor, usually appearing as a lone bird any time between late March and late September. Twelve of our thirteen sightings were of single birds and one was of three birds which obviously constituted a family party of two adults and one fully grown immature bird.

When present on the island they are restricted to the wooded areas and very rarely visit the areas west of Wadgemup Hill. The Black-faced Cuckoo-shrike mainly eats insects which they take from the foliage of trees but they will hawk for flying insects and they have been known to feed on fruits and berries.

RED-CAPPED ROBIN *Petroica goodenovii* (Plate 13).

Identification: Length 10-12 cm; weight 8.0-13.0 g. The adult male Red-capped Robin is one of the most distinctive birds on the island because of the brilliant red cap and breast. The remainder of the upper surfaces are predominantly black with a white wing bar, the under surfaces are mostly white and the tail has white edges. The eyes are brown, the bill and legs are black. The adult female lacks the red on the breast, has a rufous patch on the forehead, the upper surfaces are pale brown, some wing feathers have white edges, the tail has white edges and the under surfaces are dirty white.
Status: The Red-capped Robin is a common resident of the woodland areas, being found in all remnant *Melaleuca* woodland, most *Acacia* woodland and in some of the plantations, where they adjoin remnant woodland. This Robin is interesting because it, like the Singing Honeyeater, has probably been present since the island was cut off from the mainland about 6500 years ago. There are no Red-capped Robins on the adjacent mainland although birds do move occasionally off the Darling Scarp onto the Swan Coastal Plain adjacent to the Scarp; nor is this species present on Garden Island.

The Red-capped Robin forages at the lower levels of woodland areas, often hopping around on the ground. The birds form pair bonds and defend territories within which they nest and feed. These territories appear to be defended all year and the birds produce a low 'tinkle' call associated with their territorial defence; males chasing off other males and females dealing with intruding females.

The nest is a cup-shaped structure built in a fork of a tree and can be up to 5 m from the ground. The nest is built of bark, fibrous material and spiders' web and the female lays 2 or 3 eggs sometime between August and December. Both adults tend the eggs which are incubated for just over 2 weeks. The chicks remain in the nest for 2 or 3 weeks and stay in the territory with the adults for some time after leaving the nest while they learn to fend for themselves.

The Red-capped Robin feeds on spiders, scorpions, caterpillars, moths, flying ants and other insects. The birds may be seen on a perch close to the ground looking for prey, which when sighted, is quickly picked up. They may remain on the ground, hopping about, spreading their wings and hunching over, closing their wings and hopping forward. This may cause any prey to move and show itself, but they also turn over the ground litter in search of food. The Red-capped Robin is not strictly a ground feeder as the birds will cling to the trunks of trees, like Treecreepers, probing for food under bark; they also glean in the outer foliage and occasionally fly out to catch flying insects.

One advantage of maintaining a territory is that the occupants get to know it well and this was shown one day when one of us was sheltering under an old *Melaleuca* from extremely heavy rain. An adult female Robin was foraging in a nearby clearing during the light rain preceeding the cloudburst, but when the heavy rain poured down, the female shot off the ground directly for the heart of the clump in which DAS was sheltering. The Robin ignored DAS and flew directly for a perch under a thick, sloping *Melaleuca* trunk. During the 20 minutes the female stayed there the rain running down the trunk did not touch the spot where she was sheltering, but DAS ended up looking like a drowned rat!

Although widespread in the wooded areas, this isolated population may be small in number and genetically slightly different from its mainland relatives. The conservation of this isolate is very important and particular emphasis should be laid on the preservation of all remnant *Melaleuca* woodland on the island. The possibility of expanding these areas should also be explored to allow the Robin population to increase.

GOLDEN WHISTLER *Pachycephala pectoralis* (Plate 13).

Identification: Length 15-18 cm; weight 22.5-25.5 g. The adult male Golden Whistler with its distinctive bright yellow and contrasting black colouring is one of the most attractive birds on the island. The head is black, throat white, the chest, abdomen and back of the neck are bright yellow and the yellow is separated from the white throat by a black band. The back and rump are olive-green, the wings are predominantly dark grey, the tail is dark grey, the eyes are brown, the bill is black and the legs are grey.

The adult female and the immature birds are dull in comparison with the male; being grey on the upper surfaces and buff on the lower surfaces.

The whistlers have large heads with stout bills and it is the solid head which gives them their generic name *Pachycephala* or 'thickhead'; a rather unfortunate name for a lovely group of birds.

Status: The Golden Whistler is a resident which occurs in all remnant *Melaleuca* woodland, some of the *Acacia* woodland and some of the plantations east of Wadgemup Hill. It was formerly much more widespread on the island and Glen Storr saw them as far west as Stark Bay where they were inhabiting *Acacia* thickets inland. The wild fires of February 1955 destroyed large areas of woodland, particularly in the centre of the island and this loss of habitat led to a decrease in the range of the Golden Whistler.

On occasions, the only indication of the presence of these birds is the clear whistle-like song which often ends with a sound like a whip cracking. The birds on Rottnest Island appear to be territorial and maintain their territories all year with both birds calling to advertise the territory. On some occasions it sounds as if the male and female are engaging in a duet with one bird singing part of the song and the other completing it. The birds sing all year but are much more vocal during the breeding season between late August and early November. The nest is constructed by both birds and is cup-shaped, set in the fork of a tree and made of grass, leaves and plant stems. Two eggs are laid and incubated by both adults for just over 2 weeks.

The Golden Whistler is one of the vulnerable species on the island because it, like the Red-capped Robin, is dependent on the remnant *Malalueca* and *Acacia* woodland for its survival. Being a larger bird than the Robin, it requires a larger territory and there are fewer Whistlers than Robins on the island. No one knows what constitutes a viable size for an isolated population but every effort should be made to protect the remaining woodland and to regenerate areas of former woodland, to ensure the continued existence of those species dependent on this kind of habitat. The productivity of woodland depends on all elements of the trees including the litter created by the decay process as leaves and branches break off and form mats on the ground. These areas are rich in insects and provide food for the woodland birds, as well as aiding the break down of the vegetable matter, releasing nutrients stored in the plants and making them available for recycling.

The Golden Whistler feeds on insects and it forages in the crowns of the trees, along the branches probing under bark and it readily moves down to the ground and forages within the area shaded by the upper story of the woodland. Banding studies conducted elsewhere in Australia have shown that the Golden Whistler may live in excess of 17 years.

WESTERN WARBLER *Gerygone fusca* (Plate 13).

Identification: Length 9-10 cm; weight 5.5-6.3 g. The Western Warbler is the smallest bird occuring on Rottnest Island and it is more often heard than seen. The upper surfaces are grey-brown, the under surfaces are dirty white, there is a fine whitish line from the base of the bill to each eye, the tail has white at the base of the outer feathers and white spots near the tip of most feathers, the eyes are red and the bill and legs are black. The call of this species has a somnolent quality which gives it another common name of 'Sleepy Dick'.

Status: The Western Warbler colonised Rottnest Island in the early 1950s and quickly spread to all of the suitable habitat in the eastern half of the island. Now they are common in most of the woodland east of Wadgemup

Hill and they frequent *Melaleuca, Acacia* and plantation woodland. Their plaintive call may be heard at any time of year although they tend to call more often during the breeding season; September to December. Outside this period the call may become abbreviated. The birds appear to be territorial and it is thought that they form permanent pairs. The female builds an oval dome nest suspended from a branchlet, lays 2 or 3 eggs and incubates for nearly 2 weeks. The chicks are fed by both parents and leave the nest about 10 days after hatching.

The birds are usually seen singly, in pairs or in small family groups. Their diet consists of insects which they collect in the outer foliage of trees, sometimes hovering for quite long periods to harvest insects from a cluster of leaves.

SINGING HONEYEATER *Lichenostomus virescens* (Plate 13).

Identification: Length 19-22 cm; weight 23.0-41.0 g. The Singing Honeyeater is not aptly named as its calls are a series of 'screes' and 'chatters'; one of which sounds like a pea being blown round in a tin whistle. Both sexes appear the same; being grey-brown on the upper surfaces, streaked grey and buff on the lower surfaces, with a well defined black bar from the base of the bill through each eye to the back of the neck. Below this black bar is a thin yellow band which terminates in a dirty white patch around the ears. The eyes are brown, the stout bill is black and the legs are dark grey.
Status: The Singing Honeyeater is a common bird occurring in all habitats (with the exception of the surface of the lakes and swamps) the length and breadth of the island. Despite its size and mobility, the population has probably been present since the island was isolated from the mainland around 6500 years ago, sufficient time to enable it to evolve to the point where it is noticeably different from the population on the adjacent mainland where the species is also a common resident. The birds on the island are considerably larger than their relatives on the mainland and tend to be about 20% heavier. In fact they are so much larger than the mainland birds that they take a size 4 leg band; a size larger than that used to band Singing Honeyeaters on the mainland. They also appear to be much more heavily streaked on the lower surfaces making them seem darker than the other population, a fact which was pointed out by Lawson (Whitlock) in 1905. Scientifically the Singing Honeyeater on Rottnest Island is very important because it constitutes an endemic race of a very widespread species and is a good local example of incipient speciation.

The Singing Honeyeater is conspicuous as it forages throughout the settled areas of the island and may be seen perched on powerlines, fences and even the verandahs of cottages as it watches for prey. The birds are usually seen in small groups of ones and twos and they are extremely aggressive; both to each other and different species. They engage in noisy aerial chases and in some cases they may be so intent on fighting each other that they fall to the ground locked in combat.

The birds are scattered throughout the island but they will congregate in loose assocations when certain species flower. The birds probe in flowers for insects, although they do eat nectar, pollen and seed. Caterpillars, beetles, spiders, flying ants, flies and bugs are eaten and the birds are versatile in that they probe, hawk and glean for food using all levels of the vegetation. They have also learned to seek food in the leaf litter and debris that accumulates in the forks of trees.

Not a great deal is known about their breeding biology but there are indications that they may form permenent pair bonds. The nest is a cup-shaped structure suspended in a tree or bush, constructed of grasses and cobwebs, and is lined with soft vegetable matter or hair. The 2 or 3 eggs are laid sometime between June and December and incubation takes about 2 weeks. Both adults feed the immature birds after they leave the nest but the length of this phase of parental care is not known.

SILVEREYE *Zosterops lateralis* (Plate 13).

Identification: Length 10-13 cm; weight 7.3-12.0 g. The upper surfaces are olive-green, there is a white ring round each eye, the lower surfaces are grey with the exception of the throat and the area under the tail, near the base, which are yellow-green. The eyes are brown, the bill dark brown and the legs grey.

Status: The Silvereye or 'Greenie' is an abundant resident of the island and may be found in all habitats, including the margins of the lakes and swamps and the offshore islets and stacks. The population on Rottnest Island is probably not isolated from that on the adjacent mainland as this species is known to move over long distances (for example, Perth to Margaret River).

On the island this species may be seen in small parties or larger flocks as the birds move between food sources, calling as they fly. Silvereyes feed on nectar, insects, seed and fruit. Banding studies have shown that they may live for more than ten and a half years.

This species is probably the most numerous resident and the birds nest between August and December. The nest is cup-shaped, constructed of grasses and cobwebs and suspended in a shrub. Two to 4 eggs are laid and incubated for 10 to 12 days. After hatching the nestlings remain in the nest for about 12 days before joining the parents in the foraging flocks.

AUSTRALIAN RAVEN *Corvus coronoides* (Page 63).

Identification: Length 46-56 cm; weight 460-745 g. The corvid on Rottnest Island, often called the 'Crow' is, in fact, the Australian Raven. It is large and black with a blue sheen to the feathers when seen at close quarters. The feathers around the throat are long giving a bulging, goitrous look to the profile, the massive bill is black, the legs are black and the eyes white in adults but brown in immature birds.

The call is a drawn out 'AARK', the last syllable of the call commonly dropping in pitch as if the bird has been choked.

Status: The Australian Raven is a common, conspicuous and noisy resident. In the early days of this century it was regarded as uncommon with a few birds being seen occasionally; these were thought to be visitors from the mainland.

By the 1920s the birds were abundant in the settlement where they scavenged among the campsites. By the 1950s they had become common, especially around the settlement, the rubbish tip and the salt-lakes. They feed on dead Quokkas, rubbish, insects, eggs and young birds. It is common to see over 60 birds scattered around the rubbish tip on Forbes Hill or roosting in the *Melaleuca* woodland around Vincent or Pink Lakes. The Raven is also common around the settlements where it forages for food in and around the rubbish bins.

There is no doubt that European settlement has favoured the Raven with the provision of excess food from rubbish and scraps from picnic tables. The progressive removal of woodland and its replacement by heath has favoured the Quokka which has increased markedly as a result and this has boosted the supply of carrion.

The high population of this scavenger does pose problems but there is little point in attempting to control the birds without removing the supplementary food supply as the birds move to and from the mainland and any culled birds would soon be replaced.

The Raven forms permanent pair bonds and the birds build a cup-shaped stick nest in the fork of a tree. The remains of these nests may be seen in many trees throughout the island. Four or 5 eggs are laid between July and September and the female incubates the eggs for about 3 weeks, during which she is fed by the male. Both parents feed the chicks which leave the nest after about 6 weeks. The parents continue to maintain their young for about 6 weeks during which time the young learn to fend for themselves. After this they join the main flock and break the bonds with their parents. The immature birds do not reach maturity until they are at least 3 years old and individuals have been known to live for twelve and a half years.

SECTION V: Birds of the Heath

Nankeen Kestrel

Nankeen Kestrel *Falco cenchroides* (Plate 8).

Identification: Length 31-35 cm; with the female being larger than the male; weight 120-170 g. This is a small falcon with long wings and tail, and the ability to hover which makes it readily identifiable. The upper surfaces are rufous, with the exception of the wing quills which are brownish-black and the tail which is pale grey with a black bar near the tip in the male, and rufous with fine black bars in the female. The under sides are white with some faint brown streaks. The skin round the eyes, the cere and the legs are all yellow while the beak is blue-grey.
Status: The Nankeen Kestrel may be commonly seen on the island where it nests in the spring, using crevices in cliffs, offshore stacks or old buildings. In 1982, 1983 and 1984 a pair nested in the ceiling of an outhouse near the lighthouse on Wadjemup Hill and the birds often perched in the window recesses of the lighthouse itself. The 3-6 eggs are laid between August and October in a small depression on the floor of the nesting chamber. The nestlings are fed by both parents. On becoming independent the young birds probably leave the island for the mainland as the resident population appears to be stable at 6-10 breeding pairs.

The Nankeen Kestrel ranges all over the island, and may be seen perched on a vantage point or hovering; it is a beautiful bird to watch as it maintains station over some point on the ground, with only its head appearing to move as it scans the area for prey. On sighting some prey animal it stoops, flying off carrying it in its talons.

The diet of the Kestrel on Rottnest Island consists of skinks, legless lizards, young dugites and insects. One Kestrel was observed feeding on flying ants at the edge of the sealed road round Bickley woodland. The bird was perched on a powerline overlooking the entrance to an ant nest from which flying ants were emerging, one every few minutes. Each time an ant emerged, the bird swooped down to land by the entrance, seized the ant in its bill and swallowed it, before the ant had spread its wings. The bird then flew back to its vantage point to wait for the next victim. The Kestrel may also prey on small birds like the Welcome Swallow or White-fronted Chat.

RING-NECKED PHEASANT *Phasianus colchicus* (Plate 8).

Identification: Length 52-90 cm; weight 500-3000 g. The pheasant has a long tail and the male is slightly larger than the female and more brightly coloured. The male has an area of bare red skin around the eyes, the neck is blue with a white collar at the base. The rest of the body is reddish-brown, spotted or streaked with purple and grey-brown. The tail is grey-brown with

black bars, the bill is yellow-bone and the feet and legs grey. The female is brown, spotted or streaked with brown and grey-brown. The bill is bone, legs and feet grey.
Status: The Ring-necked Pheasant is a native of Asia Minor and has been introduced successfully to King Island (in the Bass Strait), Tasmania and Rottnest Island. This species was introduced to Rottnest Island in January 1928 when one male and three females were released. Three years later they were common all over the island being most numerous in the central and eastern parts. This is still true today with the birds being found in heath, woodland and on the edge of the lakes and swamps from Narrowneck east.

In the autumn, groups of these birds may be seen in the more open habitat and we have seen flocks of 15 foraging on the slopes of Wadjemup Hill, These flocks break up in winter with breeding taking place in spring. The female lays 8-14 eggs but little is known of the breeding behaviour.

On the island, Ring-necked Pheasants feed on snails, ants and seeds. They spend most of their life on the ground but will fly if startled. It can be alarming to be walking or riding along a track when a pheasant erupts from a bush, rising steeply with a rapid series of wing beats. Once the bird has gained sufficient altitude, it glides, with the wings outstretched, back to ground and runs off quickly.

RICHARD'S PIPIT *Anthus novaeseelandiae* (Plate 13).

Identification: Length 15-17 cm; weight 19-29 g. Upper surfaces are brown with darker brown streaking, pale buff stripe above each eye, white outer tail feathers give characteristic white stripes when the bird is flying, breast streaked with brown, eyes brown, and beak and legs pale brown. It has a habit of wagging its tail up and down when stationary and this is another diagnostic character.
Status: The Pipit is a common resident occurring throughout the heath foraging for insects which form the bulk of its diet. Early this century it was regarded as fairly common, breeding round the shores of the salt-lakes. It is now absent from the margins of the lakes but has been recorded in all of the open country the length and breadth of the island, although it tends to be seen more often in the western two thirds. The range of the Pipit has changed since the early 1900s as they disappeared from the eastern third of the island (east of line from Point Clune through the west tip of Serpentine Lake to Vera Rocks) by the 1950s but they have since re-invaded this area.

The Pipit is usually seen singly or in pairs but occasionally it may be seen in flocks of up to 6 birds. It breeds in spring, laying 2-4 eggs in a cup — shaped nest on the ground. There is no local knowledge of breeding. Nor is it know if the population on Rottnest Island is isolated from the population on the adjacent mainland.

SPOTTED SCRUB-WREN *Sericornis maculatus* (Plate 13).

Identification: Length 11-14 cm; weight 9.5-12.0 g. The upper surfaces are grey-brown, the throat and chest are buff with heavy black streaks, belly is buff, the shoulder has several black feathers which are edged with white giving a conspicuous contrasting black and white pattern, there is a prominent white stripe from the base of the bill extending above and behind the eyes, the area between the eyes and the bill is black in the male, brown in the female. The throat is pale buff with a band running roughly parallel with the stripe above the eyes. The eyes are yellowy-white, the beak brown-black and the legs pale flesh-brown.

Status: The Spotted Scrub-wren is a common resident which is found in woodland areas, including remnant *Melaleuca* and *Acacia* woodland as well as the plantations. It also occurs in small parties in the scrub heath all over the island.

This species is not often seen as it very rarely reveals itself when foraging for insects. Often the only hint of its presence may be a scolding chatter but the bird is very inquisitive and may be attracted by producing squeaking noises through pursed lips. This usually has the effect of producing a stream of calls from the Scrub-wren as it breaks cover to find out what is producing the noise.

The Spotted Scrub-wren appears to live in family parties which remain in a known area but very little is known about their breeding on Rottnest Island. Some birds banded in other locations in Australia have lived for more than 12 years.

SECTION VI: Birds of the Settlements and Disturbed Areas

Sacred Kingfisher

PEAFOWL *Pavo cristatus* (Plate 7).

Identification: Length, male 180-230 cm; female 86-100 cm; weight 2.7-6.0 kg. A large bird that spends most of its time on foot but it can fly strongly when necessary. There are distinct differences in colouration and size between the sexes. The male, which is very much larger than the female, has a royal blue neck, breast and head. The back is bright green and the upper tail coverts are greatly elongated with an eyespot near the tip. The eyespot consists of a dark blue centre surrounded by a thin band of royal blue, a larger band of copper brown and a thin band of green. As these feathers are varying lengths and they may be fanned and raised at right angles to the back, they produce a spectacular display of a fan with eyes. The underside of the body is dark green, the head has a fan-shaped crest and the bill, feet and legs are grey.

The female is drab in comparison. She has a fan-shaped crest, the head and upper neck are chestnut, and the throat and side of the head are white. The upper part of the back and breast are green while the rest of the upper surfaces are brown. The undersides are buff, and the bill and legs are light brown.

Status: This species is a native of India and Sri Lanka and the colony on Rottnest Island is the only known feral population in Australia. It is unclear when the birds were released on the island but it was done via the Zoological Gardens sometime between 1910 and 1915. They have been free-ranging since then. During the late 1950s the population was believed to be no more than 50 birds and this is probably true now (1985).

The areas in which the birds spend most of their time has changed over the years. Glen Storr found during the 1950s that the birds were mainly found round Forbes Hill in the area to the south of the wind generators and north of Lake Timperley. In fact this area is known as Peacock Flats. From here the birds ranged to Oliver and Wadgemup Hills and Mt Herschell. Since then the birds have changed their distribution markedly, shifting to the east end of the island. Until the departure of the Australian Army from Kingstown Barracks in December 1984, the bulk of the Peafowl population lived there. The Army staff fed the birds and provided them with protection. From Kingstown Barracks the birds foraged in the woodland between the barracks and Henrietta Rocks, frequently waiting for handouts from visitors near the entrance to the barracks. The birds also wandered around the settlement at Thomson Bay in small numbers and one or two birds could be seen around the old stables on the south side of Herschell Lake and around the Lighthouse Keeper's cottage on Wadgemup Hill. In fact during 1984 there was a male that remained at each of these places although it did wander some distance around. One adult male demonstrated how far they will wander, as he was banded in January 1984 and seen throughout that year. In mid-January he and another male were centred around the Lighthouse Keeper's cottage. Two days later it was seen at the turnoff to Stark Bay. Three months later it was foraging round the settlement at Thomson Bay and subsequently it remained round Kingstown Barracks.

The males are polygamous, using their spectacular tails to attract females with which to mate. These long tail feathers moult towards the end of the year and it is not uncommon to see visitors to the island cycling along clutching one of these beautiful plumes.

The male also uses the fanned tail to threaten other species and we watched a male with his tail fanned while feeding on a loaf of bread. There were several Silver Gulls standing around trying to feed on the bread. Each time one approached, the male turned so that the tail faced the gull which then backed off. The male managed to eat his fill without losing any bread to the gulls.

The 3-5 eggs are laid in a scrape on the ground sometime between October and December with the female doing all of the incubation and subsequent caring for the young. In the breeding season of 1983/84 the Army staff told us that 22 young were produced but a feral cat had eaten quite a few. They also thought that the adult birds were being taken from the island by visitors who probably fancied the idea of a bird wandering round their garden.

In addition to handouts of bread from visitors or island staff, the Peafowl feeds on seeds, insects and small lizards and snakes. The birds cause problems to the residents of Thomson Bay as they forage in their vegetable gardens.

BANDED PLOVER *Vanellus tricolor* (Plate 10).

Identification: Length 25-26 cm; weight 146-216 g. This bird is brown on the upper surfaces and white underneath. The crown is black, there is a broad white bar behind the eyes, and a broad band of black down the side of the neck and round the breast. The throat is white, surrounds of the eyes are yellow, the bill is yellow with a dark tip, there is a red wattle in front of the eyes and the legs are reddish.

Status: The Banded Plover is a bird of disturbed ground which favours the short grassy flats of the runway verge, the golf course and some lake surrounds; such as Timperley at the west end of Serpentine Lake. It also occurs on exposed sandy areas such as the sandblow at West End. This species was not found on the island in the early part of this century and the first records were of birds at Bathurst Point in 1934. From all accounts they have been resident ever since. This colonization of the island was part of an expansion of the range of this species around the Swan Coastal Plain in the 1920s.

The Banded Plover forages on the ground and its short legs restrict it to areas of low or cropped vegetation. If the vegetation gets too long the area becomes unsuitable and the birds abandon the site. This happened with the removal of horses from the island. When the stables were in operation the horses were kept in a paddock at the west end of Herschell Lake. The horses kept the vegetation short and parties of Banded Plover were often seen in the paddock.

The birds are quick to exploit suitable areas as they become available. Early in 1984, as part of the re-afforestation program, several areas were burnt and fenced. By July 1984, most of these areas had a pair of Banded Plover on them and there were signs that the birds intended to nest on those sites.

Breeding takes place in spring with 3-4 eggs being laid in a scrape on the ground sometime between July and early October. The birds will breed in any suitable site on the island. Both parents incubate the eggs and just after the eggs hatch together, the chicks leave the nest. When danger threatens they scurry for cover and 'freeze', relying on their cryptic colouration to avoid detection. If an intruder approaches the chicks, the adults will fly at the intruder calling loudly. If danger threatens the young birds or the nest, one of the adults may give a 'broken wing' display and lead the intruder away.

There are probably about 50 Banded Plover on the island and the largest group we have seen was 23 on the golf course in June. Flocks are more commonly seen in the hotter weather between November and February, but groups of two are most commonly seen. In fact nearly half of our 172 sightings of groups of birds were of a pair, with a single bird being the next most common sighting.

The Banded Plover feeds on insects and other invertebrates as well as grass seed.

SACRED KINGFISHER *Halcyon sancta* (Plate 12).

Identification: Length 19-23 cm; weight 30-52 g. This beautifully coloured bird resembles a scaled down Laughing Kookaburra *Dacelo gigas.* The upper surfaces, including the crown are greenish-blue while the lower surfaces, throat and collar are buff. There is an area of buff above and in front of the eyes and a wedge of black extending through each eye around the back of the head. The eyes are brown, the long solid bill is black and the feet are brown. Adult males have a bright blue rump while the adult females have a dull green rump, but this is hard to distinguish in the field.

Status: The Sacred Kingfisher is a conspicuous resident which is restricted to the wooded areas of the eastern half of Rottnest Island and is very rarely seen west of Wadgemup Hill. The bulk of the population lives in the wooded areas of the settlement at Thomson Bay, Bickley Woodland and the woodland round Kingstown Barracks.

During the early 1930s this species was regarded as being rare and timid but by the late 1950s and early 1960s the birds were being seen more often although they appeared not be be breeding on the island. By the 1980s the Sacred Kingfisher had established itself as a breeding species with some birds becoming permanent residents.

The Kingfisher population on the island is interesting for 2 reasons. The first is the permanent nature of their residency and the second is their choice of nest site. In the southern part of the range of this species (including the south-west of Western Australia) they are usually breeding migrants which arrive from the north in September or October. A few individuals do remain in the south but they are usually quiet and rarely seen. Our observations show that many of the birds remain throughout the year and they may be heard calling at any time of year, although the frequency of calling is considerably less during the non-breeding season.

The Sacred Kingfisher appears to be territorial, defending the area around the nest site by sitting on some vantage point (for example, the power lines over the Bowling Club) calling loudly and flying at intruding birds of the same species. One Sacred Kingfisher with a nest near the path to the Holy Trinity Church regularly sat on a signboard calling loudly. This signboard, which the bird used as a territory calling post, read 'Holy Trinity Church'. To prove that the Sacred Kingfisher is non-denominational there is another territory which includes the Rottnest Chapel. The birds remain within these territories all year and we banded two individuals in Bickley Woodland and both birds were seen several times in the same area over the following 18 months in one case and 2 years in the other. Recoveries of banded birds elsewhere in Australia have shown that they may live in excess of 8 years.

The choice of nest site is the second interesting feature of the Sacred Kingfisher on Rottnest Island. In Eastern Australia this species may excavate a tunnel in an earth bank as a nest site, but in Western Australia they usually use a hollow spout or some other hole and do not excavate their own tunnel. On the island they have selected the exotic Palm Trees as nest sites and excavate their burrow at the start of each breeding season. The staff of the Rottnest Island Board cut off the dead Palm fronds and this leaves the flat, soft, diamond-shaped base of frond exposed on the trunk of the Palm. Overhanging the trunk the live fronds provide shelter to the trunk and the nest burrows. Most of the soft-trunked Palms scattered round the settlement exhibit the remains of Kingfisher burrows in various states of disrepair. To excavate the burrow the birds fly at the nest tree and bury their stout bill into the region of the nest entrance, pulling out the pithy wood. They gradually extend the tunnel, widening it out into a chamber. The tunnel entrance may be anything from 1.6 to nearly 5 m above the ground, depending on the height of the live fronds, as the entrances are usually close to the base of these fronds. The tunnels are about 5 cm high and 6 cm wide extending into the tree for about 5 cm before widening into the nestling chamber which is about 14 cm in diameter and the roof is often domed.

The female lays 4-5 eggs some time between September and early November with both adults incubating the eggs and feeding the young. The birds feed on small reptiles, such as geckos, skinks, legless lizards and baby dugites which are common in the spring. They also feed on grasshoppers, beetles, crickets and other insects.

We have described the Sacred Kingfisher as a conspicuous resident rather than saying it is common. It has a habit of calling loudly and insistently and it sits in prominent positions. These habits, together with its spectacular colouring and its distribution on the island, taking in the settlement at Thomson Bay, makes it readily observed yet there are probably less than 25 breeding pairs on the island. This estimate is based on the distribution of 343 sightings of the Sacred Kingfisher, 85% of which were of a lone bird and all the rest (bar 1) were of 2 birds.

RAINBOW BEE-EATER *Merops ornatus* (Plate 12).

Identification: Length 19-25 cm; weight 20-31g. The Rainbow Bee-eater is a distinctive bird with brilliant plumage, pointed wings and a graceful pattern of flight. It is predominantly green with a pale green forehead, orange-yellow nape, thick black band with a thin blue edging band from the base of the bill through each eye to the back of the head, bright orange-yellow throat with a black band at the base, green chest, light blue belly, blue rump, black tail and coppery colouring on the trailing part of the upper wing surfaces. Two fine black filaments extend beyond the rest of the tail; those of the male being nearly twice as long as those of the female. The bill is long, down-curved and black, the eyes are red and the legs grey.

Status: The Rainbow Bee-eater is a regular migrant arriving from the north in early October, breeding and leaving sometime in late February or March. This species has only recently become established on the island and the first records were made by Ian Abbott and his colleagues in December 1977. Information given to Abbott suggested that this was the first year that the Bee-eater had visited the island. Since then it has been recorded each year and nests have been found at various localities from Kingstown Barracks to the turnoff to Eagle Bay.

The birds nest in a burrow which they dig in the ground, either on flat ground or into a bank. The site chosen is usually some disturbed area where the vegetation is very short or non-existent. The golf course, with its mowed vegetation, is a favoured spot and there are usually several nests scattered over it. There is always a perch somewhere within sight of the burrow and this allows the birds to scan the area before flying into the burrow. The perch also allows them to watch for food as they are insectivores, feeding on prey they catch in the air.

The nesting burrow consists of a tunnel which goes into the ground at a slope (if burrowing in flat ground). The tunnel is about 5 cm high and 6 cm wide and has 2 furrows which look like wheel ruts on the floor. These are made by the birds' feet during excavation and subsequent traffic through the tunnel. The tunnel is about 1.4 m long, the nesting chamber about 41 cm below the surface (if the burrow entrance is in level ground) and the chamber itself about 10 cm high, 22 cm wide and 24 cm long.

The 4-6 eggs are laid on the bare floor of the chamber but little is known about incubation. The Bee-eater is an interesting species because, in addition to the parents, other adults help to rear the chicks. We watched one nest, just inside the entrance to Kingstown Barracks, which had 3 large nestlings in it. Near the entrance was a large climbing frame on which the adults perched. On the occasion we observed there were 4 adults perched on the frame. Two of these were adult males, one was an adult female and we did not establish the sex of the fourth bird. We saw 3 of these birds (the 2 males and the female) fly off the frame, catch flying insects and take them down the burrow to the young. The three visits into the nest took 29, 39 and 18 seconds. At present no one knows the relationship of the helpers at the nest to the parents, but they are probably offspring from previous seasons.

As the name suggests, the Bee-eater eats bees but the birds eat other insects as well. Insect remains examined from a series of nesting chambers on the island showed that they eat bees, wasps, flies, beetles, bugs, dragonflies and grasshoppers.

After hatching the nestlings remain in the nest for about a month. After leaving the burrow they fly with the adults, being fed by the parents and the helpers as they learn to catch prey and feed themselves. At this stage of the breeding season flocks begin to form and they may be seen perched on fences and power-lines as they watch for prey. The characteristic rolling call may be heard all over the island at this stage but they are more common east of Wadgemup Hill. The largest group we have seen was of 15 birds one February, not long before the birds left the island for their trek north.

Chapter 4: Bird Watching on Rottnest Island

With 49 species of bird occurring regularly on Rottnest Island, how many can one expect to see in a trip? Obviously this will depend on time of year, experience of the observer, the distance travelled and the period spent on the island. One way to see all of the birds is to visit all of the known habitats and the purpose of this chapter is to point out those places which are most rewarding for bird-watching.

Although there are 49 species which are likely to be seen, not all are present at the same time and some are in such low numbers that they are not often seen. We have found that one can expect to record an average of 44 species over a 5-day trip. This is based on 34 visits made during all seasons between the end of 1981 and the end of 1984. The largest number of species we have recorded was 51 during a 4-day visit in early December 1982. This involved seeing 48 of the 49 listed species and 3 species of vagrant, each of one individual. These figures only include species which land on the island; they do not include species like the Australasian Gannet *Morus serrator* which may be seen flying off the coast and occasionally during the crossing to the island.

Where are the most rewarding sites to go bird watching? Most of these sites are east of Wadgemup Hill and are within riding distance of all of the settlements. They are also in walking distance, provided one has enough time. One of the attractions of the island is that all of the suitable bird-watching spots are beside established roads or tracks and therefore do not involve the observer in cross country hikes with the feeling that there is a dugite lurking under every bush. As a consequence of this accessibility, no degradation of the fragile land surfaces on the island should result from bird-watching treks. This is important because the land surfaces are very badly degraded and it takes little to damage the vegetation and start erosion by wind and water. Obviously any activity which results in further degradation is to be discouraged. Those areas which are known to be rich in number of birds and number of species are marked on Figure 6.

Anyone wishing to go on a bird walk or ride may use the maps in this book to plan to visit the various habitats of the island and spend time in each. For example, one could leave the Thomson Bay settlement via the road past Kingstown Barracks to Parker Point, Salmon Point, Salmon and Lighthouse Swamps, the central road past the wind generators, and back to the settlement via the Causeway. This trip would take an observer past remnant *Melaleuca* woodland, a lookout to some of the offshore islets (Dyer and Wallace), Watson's Glade, the Osprey nest at Salmon Point, an area of beach at Salmon Bay East (where birds of the coast often roost), some of the freshwater or brackish swamps and some of the salt-lake verges. In fact, the Causeway is one of the most rewarding areas for watching birds on the salt-lakes. This trip is about 12½ km long and takes about 2¾ hours on a bicycle. The amount of time taken does depend on how long is spent bird-watching at each vantage point. The same trip on foot takes around 4 hours.

The amount of time taken for any trip will vary for any number of reasons, but for planning purposes, it takes about 10-13 minutes for each kilometre travelled by bicycle and between 16-25 minutes on foot.

A trip to Cape Vlamingh starting and ending at the Rottnest Island Board office is 24 km long and takes about 4¼ hours by bicycle, allowing for stops to watch birds and any time between 6½ and 12 hours on foot, depending on the fitness of the hiker and the time spent watching birds. Anyone planning bird-watching trips should remember that water is scarce on the island and take enough to cover the planned trip. Those less active can make arrangements with the staff at the Rottnest Island Board office to be dropped by bus at a particular vantage point and picked up later at the same spot, or some other point, if a walk is planned. For example, one could arrange to be dropped off at Cape Vlamingh and walk to Eagle Bay, Mabel Cove and Wilson Bay. These are all good vantage points for viewing birds on the reef platforms. From there one could walk to Narrowneck and be picked up. This bus service is available only on Tuesdays and Thursdays.

There is little doubt that Rottnest Island is a relaxing place to spend some time. It caters for many different activities and there is no reason why bird-watching should not be included in the list of things people like to do. We hope that this book will answer some questions which people may ask about the birds of the island. We also hope that this book may have increased people's awareness of the environment of the island and alerted them to the fact that what for humans is a holiday resort and a great place to go, is to animals an essential part of their ecosystem.

Any book of this nature is only as good as the information on which it is based. As more is discovered about our birdlife, so this book will need revising. Anyone with any information they feel may help towards any subsequent revision is welcome to submit it to the authors.

By all means, enjoy the wildlife of the island but PLEASE DO NOT DISTURB IT. Steer clear of any nesting colonies and report any disturbance to the staff at the Rottnest Island Board. Above all, protect the environment. Have a happy holiday and enjoy your bird-watching.

Bibliography

General books about birds

Pizzey, G. (1980). A field guide to the birds of Australia. Collins: Sydney.

Reader's Digest. (1976). Complete book of Australian birds. Reader's Digest Services Pty. Ltd: Sydney.

Serventy, D.L., Serventy, V., and Warham, J. (1971). The handbook of Australian seabirds. A.H. and A.W. Reed: Sydney.

Serventy, D.L. and Whittell, H.M. (1976). Birds of Western Australia. 5th Edn. University of Western Australia Press: Perth.

Account of various research activities on Rottnest Island

Bradshaw, S.D. (editor)(1983). Research on Rottnest Island. J. Roy. Soc. Western Australia *63:* 1-61.

Articles about the birds or the environment of Rottnest Island

Abbott, I., Black, R. and Guého, N. (1978). Notes on Rainbow Birds and Fairy Terns on Rottnest Island. West. Aust. Nat. *14:* 64-65.

Burbidge, A.A. and Fuller, P.J. (1982). Banded Stilt breeding at Lake Barlee, Western Australia. Emu *82:* 212-216.

Dunlop, J.N. (1985). Reproductive periodicity in a population of Crested Terns, *Sterna bergii* Lichenstein, in South-western Australia. Aust. Wildl. Res. *12:* 95-102.

Edward, D.H.D. (1983). Inland waters of Rottnest Island. J. Roy. Soc. W. Aust. *66:* 41-47.

Frith, H.J. (1967). Waterfowl in Australia. Angus and Robertson: Sydney.

Frith, H.J. (1982). Pigeons and Doves of Australia. Rigby: Adelaide.

Holsworth, W.N. (1965). Nesting success of the Osprey on Rottnest Island, Western Australia. West. Aust. Nat. *10:* 13-15.

Iveson, J.B. and Hart, R.P. (1983). *Salmonella* on Rottnest Island: implications for public health and wildlife management. J. Roy. Soc. W. Aust. *66:* 15-20.

Lawson, F. (1905). A visit to Rottnest Island, W.A. Emu *4:* 129-132.

Pen, L.J. and Green, J.W. (1983). Botanical exploration and vegetational changes on Rottnest Island. J. Roy. Soc. W. Aust. *66:* 20-24.

Playford, P.E. (1983). Geological research on Rottnest Island. J. Roy. Soc. W. Aust. *66:* 10-15.

Minton, C.D.T. and Lane, B.A. (1984). Counting and banding waders in Australasia. Pp. 43-54. In: Methods of censusing birds in Australia. (Ed. S.J.J.F. Davies). Dept. of Conservation and Environment. Bulletin No. 153: Perth.

Nicholls, C.A. (1974). Double-brooding in a Western Australian population of the Silver Gull, *Larus novaehollandiae* Stephens. Aust. J. Zool. *22:* 63-70.

Reid, D.R. (1950). Nesting of the Pied Cormorant off Rottnest Island. West. Aust. Nat. *2:* 69.

Riggert, T.L. (1977). The biology of the Mountain Duck on Rottnest Island, Western Australia. Wildlife Monographs *52:* 1-67.

Saunders, D.A., de Rebeira, C.P., and Smith, G.T. (1984). Birds of Rottnest Island — a checklist. Second (revised) edition (privately published: Guildford, Western Australia).

Serventy, V.N. (1950). Fairy Terns on Rottnest Island. West. Aust. Nat. *2:* 126-128.

Storr, G.M. (1964). The avifauna of Rottnest Island, Western Australia. I. Marine birds. Emu *64:* 48-60.

Storr, G.M. (1965a). The avifauna of Rottnest Island, Western Australia. II. Lake and littoral birds. Emu *64:* 105-113.

Storr, G.M. (1965b). The avifauna of Rottnest Island. III. Land birds. Emu *64:* 172-180.

Williams, C.K. (1979). Ecology of Australian Chats (*Epthianura* Gould): reproduction in aridity. Aust. J. Zool. *27:* 213-229.

Wooller, R.D. and Dunlop, J.N. (1979). Multiple laying by the Silver Gull, *Larus novaehollandiae* Stephens, on Carnac Island, Western Australia. Aust. Wildl. Res. *6:* 325-335.

Wooller, R.D. and Dunlop, J.N. (1981). Annual variation in the clutch and egg sizes of Silver Gulls, *Larus novaehollandiae.* Aust. Wildl. Res. *8:* 431-433.

Appendix 1. List and status of birds recorded on or over Rottnest Island. Sea birds which do not land on the island have not been included.

Common Name	Status	Notes
Australasian Little Grebe	Vagrant	
Hoary-headed Grebe	Vagrant	
Little Penguin	Vagrant	
Rockhopper Penguin	Vagrant	
Little Shearwater	Once bred on Island. Vagrant	
Wedge-tailed Shearwater	Breeding migrant	
Darter	Vagrant	
Little Pied Cormorant	Resident	
Pied Cormorant	Resident	
Black Cormorant	Vagrant	
Little Black Cormorant	Vagrant	
Red-tailed Tropic Bird	One breeding record. Vagrant	
White-faced Heron	Vagrant	
Large Egret	Vagrant	
Reef Heron	Resident	
Little Bittern	Vagrant	
Straw-necked Ibis	Vagrant	
Black Swan	Resident	
Mountain Duck	Resident	
Black Duck	Non-breeding migrant	
Grey Teal	Non-breeding migrant	

Common Name	Status	Notes
Wood Duck	Vagrant	
White-eyed Duck	Vagrant	
Musk Duck	Vagrant	
Osprey	Resident	
Black-shouldered Kite	Vagrant	
Brown Goshawk	Vagrant	
White-breasted Sea Eagle	Vagrant	
Spotted Harrier	Vagrant	
Little Falcon	Vagrant	
Brown Falcon	Vagrant	
Nankeen Kestrel	Resident	
Peafowl	Resident	
Ring-necked Pheasant	Resident	
Pied Oystercatcher	Resident	
Sooty Oystercatcher	Vagrant	
Banded Plover	Resident	
Eastern Golden Plover	Vagrant	
Grey Plover	Non-breeding migrant	
Hooded Dotterel	Vagrant	
Large-billed Dotterel	Vagrant	
Red-capped Dotterel	Resident	
Mongolian Sand Dotterel	Vagrant	
Black-fronted Dotterel	Vagrant	
Pied Stilt	Vagrant	
Red-necked Avocet	Resident	

Common Name	Status	Notes
Banded Stilt	Non-breeding migrant	
Ruddy Turnstone	Non-breeding migrant	
Whimbrel	Non-breeding migrant	
Grey-tailed Tattler	Non-breeding migrant	
Greenshank	Vagrant	
Common Sandpiper	Vagrant	
Terek Sandpiper	Vagrant	
Bar-tailed Godwit	Non-breeding migrant	
Knot	Vagrant	
Sharp-tailed Sandpiper	Vagrant	
Red-necked Stint	Non-breeding migrant	
Curlew Sandpiper	Non-breeding migrant	
Sanderling	Non-breeding migrant	
Northern Phalarope	Vagrant	
Silver Gull	Resident	
Pacific Gull	Vagrant	
Caspian Tern	Resident	
Whiskered Tern	Vagrant	
White-winged Black Tern	Vagrant	
Bridled Tern	Breeding migrant	
Roseate Tern	Vagrant	
Crested Tern	Resident	
Fairy Tern	Breeding migrant	
Lesser Crested Tern	Vagrant	
Domestic Pigeon	Vagrant	

Common Name	Status	Notes
Spotted Turtledove	Resident	
Laughing Turtledove	Resident	
Brush Bronzewing	Extinct	
Galah	Vagrant	
Purple-crowned Lorikeet	Vagrant	
Port Lincoln Parrot	Vagrant	
Rock Parrot	Resident	
Pallid Cuckoo	Vagrant	
Fan-tailed Cuckoo	Breeding migrant	
Horsefield's Bronze Cuckoo	Vagrant	
Golden Bronze Cuckoo	Vagrant	
Boobook Owl	Vagrant	
Fork-tailed Swift	Vagrant	
Sacred Kingfisher	Resident	
Rainbow Bee-eater	Breeding migrant	
Welcome Swallow	Resident	
Tree Martin	Resident	
Richard's Pipit	Resident	
Black-faced Cuckoo-shrike	Vagrant	
White-winged Triller	Vagrant	
Red-capped Robin	Resident	
Rufous Whistler	Extinct	
Golden Whistler	Resident	
Grey Fantail	Vagrant	
Willie Wagtail	Vagrant	

Common Name	Status	Notes
Spotted Scrub-wren	Resident	
Western Warbler	Resident	
Red-wattle Bird	Vagrant	
Singing Honeyeater	Resident	
White-fronted Chat	Resident	
Striated Pardalote	Vagrant	
Silvereye	Resident	
Australian Magpie-lark	Vagrant	
Magpie	Vagrant	
Australian Raven	Resident	

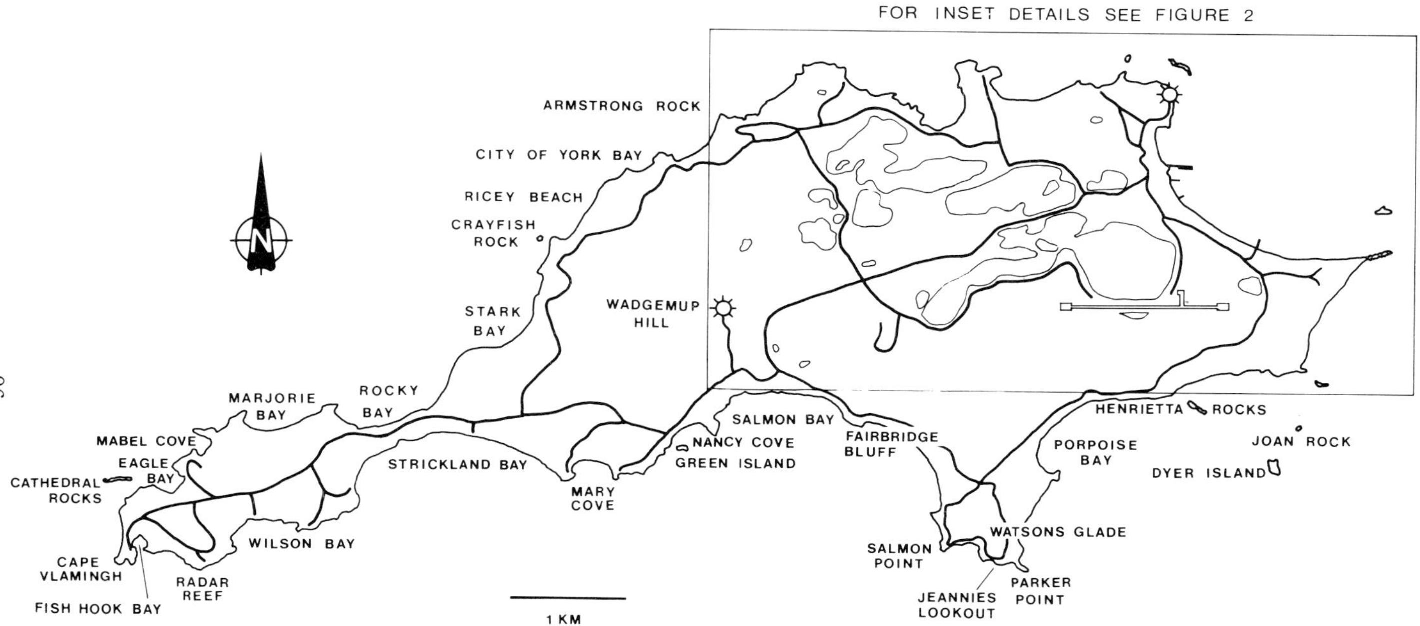

Figure 1. Map of Rottnest Island showing salient features and places mentioned in the text.

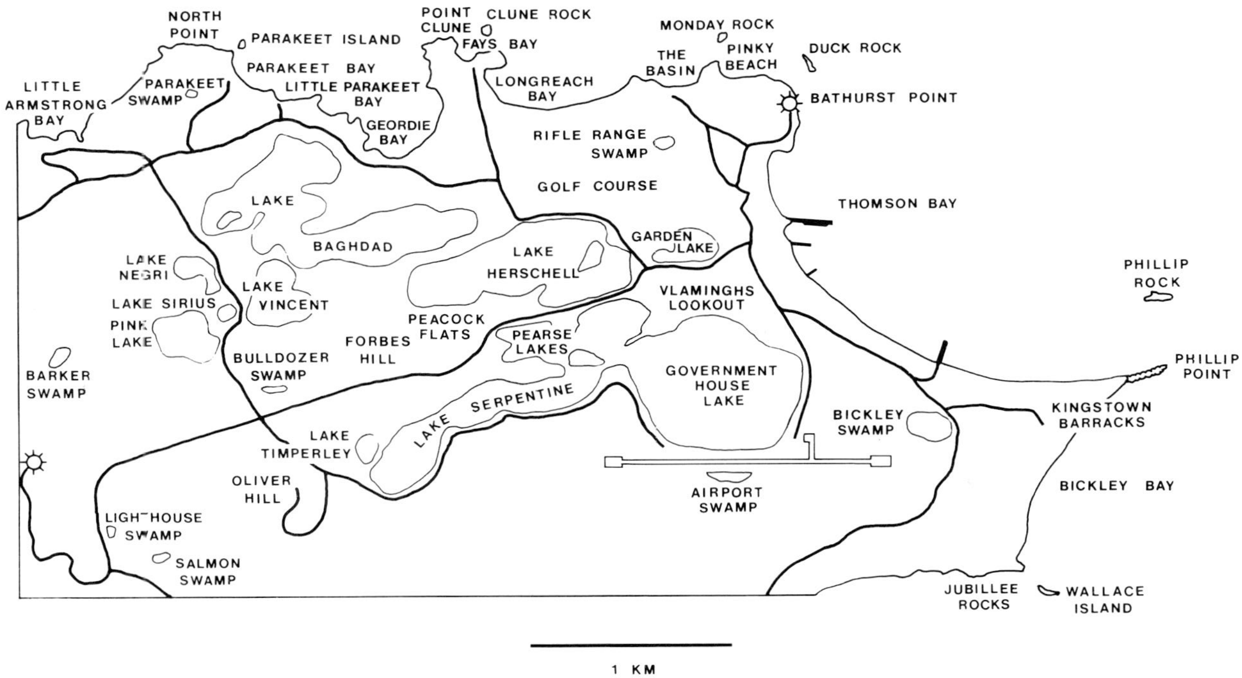

Figure 2. Map of salt-lakes and brackish swamps.

Figure 3. Map of vegetation associations and habitat types.

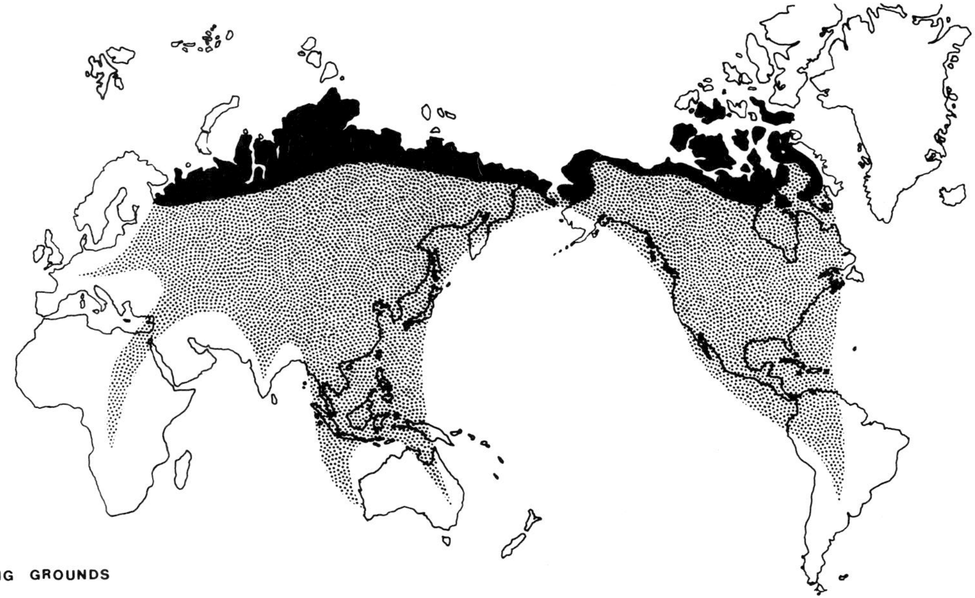

Figure 4 Map of the World to show migration routes and breeding grounds of transequatorial migrants.

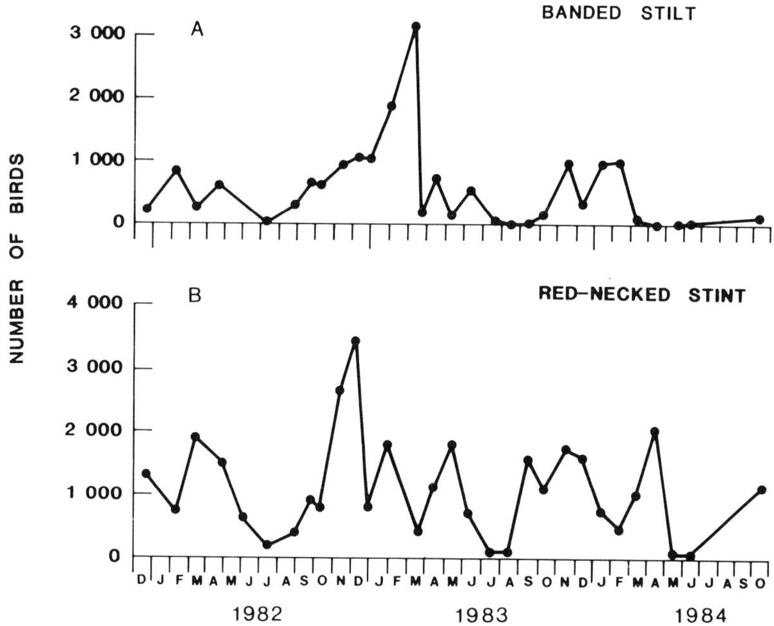

Figure 5. Graph of changing numbers of A) Banded Stilt and B) Red-necked Stint to show pattern of occurrence of a local migrant and a transequatorial migrant.

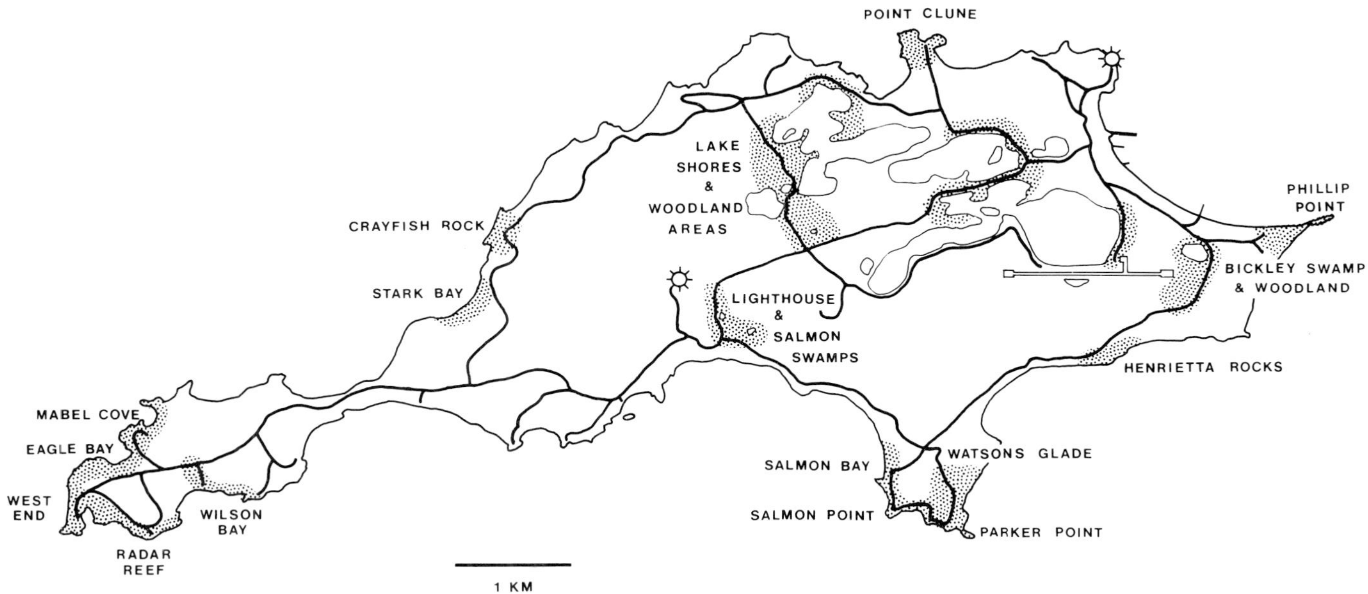

Figure 6. Map of Rottnest Island to show recommended vantage points for watching birds.